Ray Mears'
World of
Survival

Ray Mears'
World of
Survival

By Ray Mears and Jane Hunter

HarperCollins*Publishers*
77-85 Fulham Palace Road
Hammersmith
London W6 8JB

First Published in Great Britain by
HarperCollins*Publishers* 1997

10 9 8 7 6 5 4 3 2 1

Text © Ray Mears and Jane Hunter 1997
Photographs © Ray Mears

ISBN 0 00 472083 0

Editors: Myles Archibald, Ian Drury
and Liz Bourne
Designers: Clare Baggaley and Julie Francis
Production Manager: Bridget Scanlon

Colour reproduction: Saxon Photolitho,
Norwich, England
Printed and Bound by: Bath Press Colour
Books, Glasgow

▲ **Off Baffin Island, Canada, the ice can
be over two metres thick, but the
Inuit dig through it to fish for Arctic char.**

Dedication
**To the tribal elders everywhere who are teaching their youth to maintain
their traditions.**

We live in an age when many people wish to return to simpler more
natural lifestyles. Native beadwork and design have become
talismen for ancient philosophies that are held up as a shining
direction back to nature. But how many people today really know
how it is to live one on one with nature? Borne out of the success of
the BBC2 series 'Tracks', I was given the opportunity to visit six
communities living close to nature to showcase their lifestyle and
the techniques that they employ to live with nature.

The presence of the camera brought a range of differing
responses. In Siberia, children watched rushes played back on a
monitor without recognising that they were watching themselves.
On Savaii, Western Samoa, the villagers were convinced that we
were trying to slow them down when asked to repeat an action for
the camera and consequently did things faster and faster. I can
appreciate their frustration. Seen from a layman's point of view, film
making is a painfully slow process with much more film shot than
actually ends up being transmitted. But that is not all that goes
unseen. Behind every television programme there is an army of
skilled technicians that make it happen. The making of the series on
which this book is based involved an incredible number of
resources and was a major task in planning and co-ordination. With
each episode having to be filmed in
distinctly different remote environments,
from -40°C in the Arctic to approaching 60°C
in Africa, it was a major task through to its
conclusion, no matter what the odds or the
level of fatigue. As with most of the successful
expeditions that I can recall, our team was a
small one comprising two parts: the crew
that actually went into the field and did the
filming, and the production team that made it
happen, managing the life support of the film
crew and then polishing the film for
transmission. Without the full participation of
all those involved, the film would not have
been possible. My sincere thanks go to all of
them. A few individuals deserve particular
note.

Kathryn Moore, the series producer, who
bore the responsibility throughout. A
thankless and frustrating task, most
particularly when the crew was in remote
locations. Kathryn's vision and dedication was critical to the
completion of the project.

Barry Foster and Alan Duxbury, the series cameramen, who by
my reckoning have the hardest job in the field. Sharing igloos and
tepees, you soon get to know people; no finer fellows could ever be
found to join an expedition. Barry thrived on the challenge of filming

in the greatest extremes, keeping his camera running despite frozen fingers on Baffin Island, and the desperate heat of the shadeless Namibian dry season. Alan was in his element lighting the fire in the Siberian chum each morning and playing, or more accurately losing, at cards with the Evenk in the evening. Both men have an irrepressible talent for telling terrible jokes that was a welcome distraction from the merciless bites of mosquitoes.

Sam Cox, Andrew Morton and Graham Hoyland, all sound men in every sense of the term, recorded the words and atmosphere to accompany the pictures. The least obvious of all input to the films, yet without their attention to detail and alertness to sound problems the films simply would not work. Asking people who have never seen a television to be quiet during filming can be a delicate and frustrating task. They managed it with tact and sympathy, winning friends in the process. Sam doubled as an electrician and mechanic, in Australia stripping down a damaged monitor and repairing a broken generator. Difficult tasks at the best of times, but even more difficult when working by torchlight and being bitten by mosquitoes.

Dennis Jarvis, Joe Ahearne, Ben Warwick, Ian Paul and John-Paul Davidson directed the films. A difficult challenge that they each responded to in their own particular ways, bringing individual flair to each episode.

Jane Hunter who co-ordinated the logistics for the films and produced the text within these pages.

Namibia has a drought severe by its own harsh standards. People and animals struggle to survive in temperatures of up to 60°C.

Polynesian navigators discovered western Samoa thousands of years ago. The sea sustains these communities, but can also destroy them in an instant: wholesale changes to the islands' ecology have followed some of the great typhoons.

Contents

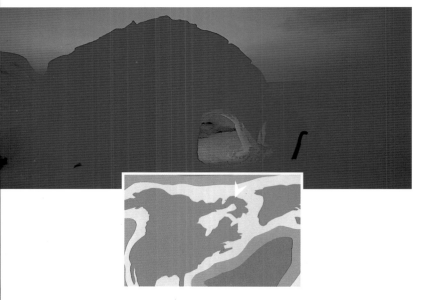

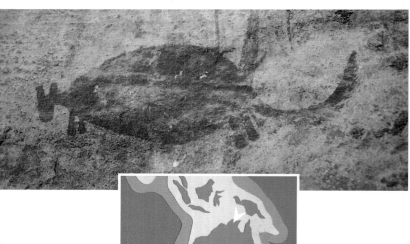

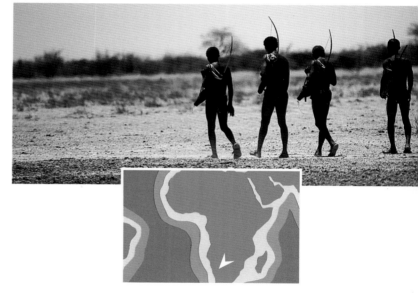

Baffin

The great paradox of the Arctic is that while water provides both the surface underfoot and can be made into housing, the land is actually a desert. Obtaining drinking water in this frozen environment remains a key survival skill.

Island

Baffin Island

An everchanging wilderness of white and blue horizons stretching in all directions, Baffin Island invites the human spirit to expand on journeys across them. At first glimpse it seems incredible that any human being can live here at all, and yet the Inuit do. With skills learned over centuries they epitomise the adaptability of our species. Turning the most unlikely of resources to their aid, they have solved the problems of living in a freezer.

Throughout its lengthy winter the Arctic fulfils its image of a wind-swept, snow-covered and inhospitable landscape. With temperatures dropping to extremes of -45°C the seas and lakes are permanently frozen for nearly nine months of the year and during winter it is plunged into twenty four hours of darkness. There is snow and ice everywhere, but water is scarce making the land a virtual desert. Yet man has survived here for several thousand years not only existing with, but depending upon these inhospitable conditions. Life in the Arctic for the Inuit people has been subjected to great change over the past three decades, but their ancient skills have not all been lost to modern technology. Many of the skills that ensured their existence in the past are still being practised and depended upon today.

The ability to find shelter from adverse weather is a fundamental survival skill. On the frozen sea around Baffin Island this need is at its greatest. The Inuit solution is perhaps the most symbolic of all shelters; the igloo. The real lesson here is in the way that they have turned the most abundant resource available into a building material.

The Place

Baffin Island, the world's fifth largest island, is located in the north east Canadian Arctic and covers an area of 1.9 million kms² of mountains, glaciers, tundra and ocean. An area twice the size of the UK, the island is 1000 miles long with a northern shoreline that is deeply indented by bays, inlets, cliffs and spectacular fjords. Lancaster Sound on the north coast, supports one of the richest populations of marine mammals in the Arctic. Some parts of the south west coast were explored and mapped as late as the 1930s and vast tracts of the island still remain unknown and unnamed.

Although the brief summers bring some life and colour to the landscape, temperatures even then rise to only just above freezing and snow flurries can occur at any time. The days are long with a period of 'midnight sun', twenty four hours of daylight. Under the ground surface lies the permafrost, a layer of frozen soil and rock which can be up to 700m in depth. Above the permafrost, only a few inches of soil will thaw each spring to allow a few, highly specialised plants to grow. This peaty soil supports acid-loving heath plants such as bilberry, cranberry and heather-like shrubs as well as grasses, lichens and mosses. The fierce Arctic winds allow a few trees to grow, but those that do remain small. The impermeable permafrost layer prevents drainage of the summer melt waters and therefore the land surface is a mass of ponds and bogs; an ideal environment for mosquitoes and blackflies.

The People

The indigenous inhabitants of the Canadian Arctic are known as the Inuit, the term they use for themselves. They belong to a linguistic stock termed Eskimo-Aleut (or Eskaleut) named for its two major branches, Aleut and Eskimo. The larger branch, Eskimo, divides near the Bering Straits, with the Yupik on one side comprising at least five linguistic groups in eastern Siberia and central and southern Alaska. On the other, the Inuit extending from northern Alaska to Greenland, including all of Arctic Canada.

In Canada the word Inuit has almost totally replaced Eskimo. The word Eskimo is thought to have derived from a derogatory Algonkian term meaning 'eaters of raw meat'. But, not all Eskimos are Inuit. Inuit differ from other native Canadians in their physical features as they are markedly Asiatic. Throughout their distribution, the Inuit speak a single language with each group

▼ To us the igloo maybe the symbol of ancient Inuit survival skill. But today only the older hunters use them, the younger men preferring to use quickly erected tents and snow machines. Their pragmatic adaptability to new ideas is perhaps an equally important Inuit lesson in survival.

How to build an Igloo

having its own dialect. These distinctive groups, small communities of hunters and trappers, would have been separated by vast tracts of Arctic wilderness, each with its own social and material culture. Some of the original groups of northern Baffin Island include the Anaulirialingmiut or 'Fish Clubbing Place People, the Natsilirjuut, 'People where there are Seals' and Kuuganajurmiut, the 'Large River People'. The resources available to these people were always limited and in Arctic Canada more so than any other Arctic region. The Inuit and their predecessors survived on whatever resources were available, basing their economy on a combination of fishing and hunting for both land and sea mammals on the ice in winter and from kayaks in summer.

Inuit society is based on equality and sharing. Originally they did not worship gods but believed that everything which moved possessed a spirit, including all the hunted animals and birds. The success of a hunt rested on having a proper attitude towards the prey animal and when caught the hunter would ask for the animal's spirit to be returned to the wild. To communicate with the spirit world the services of an angakkuq, or shaman, would be used and belief extended to many spirits such as Sila, the force that controls and oversees everthing that people do.

◆ Ham Kadloo and Jacob are elders in the Pond Inlet community. The last generation to have grown up in igloos, they built them as a daily necessity while on hunting trips. Their hands carve snow blocks with a deft ease that comes not just from practise, but from the reality of having lived with this skill. Younger listen respectfully to their advice; respect for elders is part of the Inuit way of survival.

Inuit Prehistory

Archæologists believe that man has occupied the Arctic Circle for over 4000 years in a number of stages. An Eskimo group, known as the Dorset Culture, were present in the eastern Arctic from around 800BC. They were seasonally nomadic, making delicate carvings from ivory and antler bone and living in small rectangular, partly submerged houses. They survived until about 1300AD and the Inuit today still

◆ Building an igloo is not as easy as it may seem. The snow must be of just the right strength. Blocks are quarried and used to build a spherical wall. The secret of the igloo's strength lies in the way the wall spirals upwards: each block is locked in place by gravity. As the blocks are laid they are fitted so that three points are in contact. At these points the snow crystals mix and weld the blocks together. The top block is fitted last, locking the igloo together. After an hour or so the igloo is strong enough for its builder to stand on top.

speak vividly of these people whom they called Tuniit. Between 900AD and 1300AD they disappeared as a wave of immigrants from Alaska displaced them. The newcomers were the Thule people who migrated along the Arctic coastline. This culture, considered classically Eskimo, survived until about 1750AD. Between 1650AD and 1850AD there were quite dramatic climate changes and the Little Ice Age forced people to withdraw from the high Arctic. As the climate cooled, people became more nomadic and could no longer rely on local hunting which marked the end of the Thule Culture. Sites belonging to all these cultures can be found on Baffin Island.

William Baffin

Baffin Island is named after the English explorer, William Baffin (1584-1622). He mapped Baffin Island during a series of voyages in search of the Northwest Passage. In 1615, as pilot of the *Discovery*, sailing under Robert Bylot to Hudson Bay, he obtained the first longitude to have been accurately measured at sea. It was the search for the fabled Northwest Passage (a navigable sea route from the Atlantic to the Pacific) that brought a succession of European explorers to Baffin Island and its surrounding waters. Trade with Asia promised fabulous riches for European companies, but the implaccably

hostile Ottoman Empire blocked the traditional routes through the Mediterranean and Middle East. English explorers led the way northwest, into the Arctic where they discovered that during the brief summer, the ice melted sufficiently for sailing ships to enter the waters between Greenland and Baffin Island. The Inuit's first contact with Europeans took place in 1576 when the great Elizabethan seafarer Martin Frobisher braved mountainous seas

Traditionally the igloo was heated with a Koodlik seal fat lantern or stove, keeping the internal temperature to around 0°C. That may seem chilly but when the external temperature is -35°C or lower it is a great comfort.

A hunter maintains his lonely vigil beside a seal's breathing hole. If necessary he will wait absolutely motionless for up to three hours. Rifles have made this hunting easier but the ancient skill of knowing which breathing hole to wait by survives. Note the harpoon, nowadays made of metal, which is essential for retrieving the dead seal.

Seal Hunting

off Greenland to reach the south-eastern shore of Baffin Island. Relations varied: several of Frobisher's men were killed by one party of Inuit, but the seamen traded with another group and even taught them the rudiments of football. Frobisher was an outstanding seaman, but his discovery of 'Fools Gold' (iron pyrites) on Baffin Island ultimately bankrupted the Company of Cathay that financed these voyages. He returned to Baffin Island in 1576 and 1577, mining for gold around Frobisher Bay in the south-east of the island. Evidence of his brief visits has been discovered on Warwick island.

The Search for Franklin

The search for the Northwest Passage ceased in the mid-seventeenth century after several expeditions ended in catastrophic loss of life. Ignorant of both Inuit techniques and dietary requirements, overwintering took a fearful toll. Similar problems overshadowed the next expeditions to the area, launched in the early 19th century. The most famous casualty was Admiral Sir William Franklin who perished with all 134 men in his ill-fated voyage, 1845-7. The disappearence of his two ships triggered an intensive search operation: over a dozen ships searching the Canadian Arctic over the next five years. None were able 'to reach the hand of Franklin', and in 1854 skeletons were found on King William Island. Inuit tales of starving Europeans, and further grisly discoveries combine to create a historical mystery that is still being unravelled. Ironically, the search for Franklin finally did locate a Northwest passage: it was barely navigable and, ironically, no longer of any commercial importance.

By the 19th century commercial whaling had come to the Canadian Arctic, moving from west to east in their exhaustive search for whale products. Baffin Island provided oil for the lamps of Europe and whalebone (baleen) for the corsets of Victorian women. British whaling vessels arrived at Ponds Bay, now known as Pond Inlet, in the north east of Baffin Island in the 1820s, searching in particular for bowhead whales. Subsequent trade with the local people began to introduce rifles, steel knives, sewing needles, tobacco and tea into the Inuit culture. The Inuit also hunted the animals that fed on the whale carcasses and salvaged wood from the whalers vessels. After the decline of the whaling industry around the turn of the century, traders and pioneers began to settle in the area trading southern items for seal skins, fox and bear hides and ivory tusks. The Hudson Bay Company established a trading post and both Anglican and Roman Catholic missionaries had arrived by 1929. Some became truly part of the communities, learning the language and marrying into Inuit families.

Dog Teams

Arctic Living Conditions

With the January air temperature on a still day, without the wind chill factor, between -24°C and -36°C, and average temperatures rising to above freezing for only three months of the year, June, July and August, Arctic societies needed to depend upon and not simply exist in defiance of these cold conditions. Each Inuit culture had its own ways of taking advantage of winter. As the growing of crops is impossible due to the frozen terrain, the traditional Inuit diet consisted of almost exclusively meat, fat and fish. Their lives in the past revolved around the search for food, a diet of seal, caribou, walrus, polar bear, musk oxen, whale, arctic hare, fish and birds. There are a number of advantages that exist within this 'winter' environment, not least that animal tracks are at their clearest in fresh snow and seals are particularly vulnerable to harpooning when taking air through breathing holes in the ice. Protection and warmth were essential and this is still provided by clothing made from caribou skin which can insulate against most weather conditions; two layers of caribou hide will allow a person to sleep in the open in temperatures as low as -30°C.

Shelter

It is a well quoted myth that the Inuit lived permanently in snow houses called igloos or iglus. For most Inuit these snow houses were nothing more than a temporary hunting shelter. In their language the word for a snow house is illuviga, whilst iglu means house of any kind. Their traditional winter homes were built half underground and were made of rock and clumps of earth; whale bones were often used as rafters and skins for insulation. In summer they lived in tents made from seal or caribou skins. Living now in houses with modern amenities, igloos are still built as overnight shelters on hunting trips in winter, whilst tents are used in the summer. It takes three skilled men less than an hour to build an igloo, one cuts the blocks from the packed snow, another builds and the third will plug the holes. The size is dictated by the tallest man, who lies down on the snow with his arms outstretched above his head. This will give a diameter large enough to accommodate all the men and their equipment. If they stay for more than one night, they will build a platform to lift them off the floor and away from the sinking cold air.

Snow scooters are now the most common form of Inuit transport, but dog teams are still in use pulling sledges that have changed little for centuries. The elders reflect that hunting caribou with a dog team is more difficult than with a snow machine which frightens the caribou less, but that on long journeys a machine can break down or run out of fuel, leaving you stranded. At least with a dog team, you can always eat the dogs in an emergency.

Making Water

Caribou Hunting

The caribou was the most important land mammal, providing meat, hides for clothing, sinew for thread and antlers for tools. They were traditionally hunted with bow and arrow made from antler, bone and muscle fibre, or speared from kayaks as they swam across lakes and rivers. Still an important food source, now a hunter will only kill enough for a week's supply of meat for his family, though in the past as many animals as possible would have been killed to feed the whole community.

The caribou are butchered on the ice with all parts of the animal being utilised. It is the sinew on the backbone that is used for thread, after being cleaned and dried, and it is so strong that it will outlive the fur clothing. In less than half an hour, caribou antlers can be turned into an axe which is used to strip down seal fat to make a lamp and the lower jaw and head could be made into a sledge for children.

Seal Hunting

Seals provided food for humans and dogs, oil to fuel lamps and fires, hides for boots, summer clothing, tents, harpoon lines and dog harnesses. Seals were hunted with harpoons, essential for hunting sea mammals, and many forms of harpoon were developed for different species and various hunting techniques. Seal hunting demands great patience, as a hunter often has to wait motionless over a seal hole for many hours, sheltered only by a low wall of snow-blocks. As seals need to take a breath once every half an hour, they rise to the surface using holes about 50cms wide that they make in the ice. As winter progresses, the ice thickens and forms over the holes so the seals rise up and break through the ice to reach the air. The ice pattern, and the smell around the hole, will tell the hunter how old the hole is and how recently it has been used.

With so little fruit or vegetables in their diet, seal meat was often eaten raw which provided the Inuit with their main source of vitamin C. The seal's skin is used for clothing and makes the most effective windproof trousers and gloves as it is thin, tough

Dehydration in this environment can hasten chilling and the onset of frostbite. Every Arctic traveller must know how to find drinkable water in times of emergency. Here, Jacob cuts a ladder of small depressions in the surface of an iceberg, linking each with a small channel filled with clean snow to make a water purifying system.

A fire is lit with whatever is available, in this case an old sledge board and seal fat. A block of crystal clear fresh iceberg ice is propped above the fire. As this melts the water trickles under the fire into the ladder of depressions. The water carries fats and oils from the fire which are gradually filtered out by the snow in the linking channels. The clean water must be drunk immediately or it will freeze.

Cooking on a Koodlik

and water resistant, particularly to salt water. The fineness of the seal material depends on the woman making the cloth and traditionally she would have made clothing under the guidance of the hunter and wearer.

The seal blubber, which is scraped from the skins, was stored in bags which in time produced the fuel oil for lamps and fires. The soapstone lamps, or koodlik, would provide heat and light for the home, cook the food, melt water and dry clothing.

Like the caribou, all parts of the seal are used and nothing wasted, and the Inuit are now regularly used to train the military in Arctic survival techniques to include the effective use of all parts of a hunted animal. Seal, like caribou, is still eaten regularly. Other prey animals would have included the musk ox which were easy to hunt due to their habit of standing in defensive circles when attacked. Polar bears were hunted in the past for meat and hides and a number of birds including waterfowl and ptarmigan. The walrus provided meat and blubber and ivory from the tusks and its tough hide was used for covering boats. The smaller whale species, such as the narwhal and beluga, were also hunted around Baffin Island.

◭ **Although the Coleman pressure stove is to be found in every Inuit sledge box, the traditional koodlik is still commonly used. A more reliable stove would be hard to find, and its warming glow is a welcome sight in the igloo. A kettle can be suspended over the flame and traditionally a netting rack would also be set above that to dry out damp clothing.**

Fishing

The Inuit never used fishing hooks, instead they speared fish either from the ice edge, from kayaks or, in winter through ice holes. Today, two methods are used, spearing and gill netting.

In winter the ice can be up to two and a half metres thick which requires a great deal of hard work to break through it. In the Pond Inlet area fisherman still use a wooden digging stick with a sharp metal point, with only a handful of people resorting to using electric drills. Using a shiny lure dangled down through the hole, which would have been made of something shiny like walrus ivory, the fisherman hopes to attract the fish, maybe an arctic char. As the fish approaches the lure it is speared.

Clothing

The various items of caribou and seal skin clothing are all still prepared by the women. Despite the recent progress in producing man-made extreme cold weather clothing, it is still not as effective as caribou skin. Military recruits, winter training in the Arctic, have now reverted back to it as the only answer to coping with temperatures as low as -50°C. Down clothing may be warm but it is not breathable and sweat freezes on the inside; sweat passes through caribou skin and then freezes on the outside. Only the summer caribou coats are used to make winter clothing as the caribou's winter coat is too thick. The winter coats would have been used for padding or for packing quarry, maybe seals onto a sledge.

Traditionally the women made clothing in small unheated huts, because if you take caribou fur into the warm, the hair will fall out and it tends to produce an unpleasant smell. The process involves cleaning and tanning the skins before cutting and constructing the items using caribou sinew for thread and originally bone needles. It takes approximately three days to complete a caribou parka.

Food Preparation

After a kill the meat and fat of an animal is either eaten raw or boiled, with little waste. Meat can always be stored outside due to the cold, though now freezers tend to be used which prevents the meat getting burnt by the sun, thereby preserving the taste.

Most meat was eaten raw not only to provide essential minerals, but also due to the shortage of fuel. The usual method was to bite into a strip of meat and then cut it off near the lips with a knife. Land and sea food, seal and fish for example, would never be eaten together. A nomadic group of Inuit would attempt to provide for all its members so that the sharing of all food was essential to the group's survival. But they needed to continue moving to find these food sources and if elderly or less able members could no longer keep up with the group, they would have been abandoned. The elderly and young would be the first to perish if food ran out as it sometimes did in late winter and starvation of whole Inuit groups was not uncommon.

Despite the availability of modern and convenience foods now in Inuit communities, many families still eat a high percentage of wild meat and 'land food', as they term hunted caribou, fish or seal. It is still much preferred to bought, or 'shop food'.

Obtaining Water

The paradox of the Arctic is that despite being surrounded by snow and ice, it is a virtual desert. Obtaining drinking water was a specialised skill with the lack of fuel or handy Coleman stoves to melt either snow or ice. One of the methods used was to cut out the stomach of a newly killed caribou, clean it and then pack it with fresh snow. It would then be placed back inside the body cavity to allow the body heat to produce water.

How to make a
Koodlik

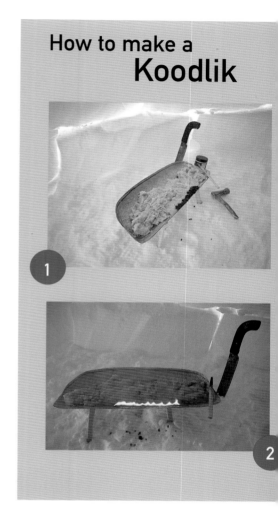

1 Raised above the floor by three sticks, it is filled with seal fat that has been pounded soft with a caribou antler hammer, seal fat is so rich in oil that the pounding is enough to liberate sufficient oil to start the lamp burning.

2 The wick is made from moss and the downy seed head of cotton grass, collected in the brief Arctic summer. The wick, once trimmed correctly, will burn evenly and steadily for many hours, providing warmth to melt the seal fat which feeds the wick.

Scraping Skins

Dogs

Dogs were extremely important to the Inuit, but their drain on food resources meant they kept only the number they needed. They were invaluable in hunting, sniffing out the seals' breathing holes under the snow and holding musk ox and polar bears at bay. In summer they were used as pack animals and in winter to pull sledges. The snow and frozen seas made travel by dog sledge over long distances possible. The sledges were made of crossbars which were lashed with sealskin thongs to the runners so they were extremely flexible. The runners were covered with mud or moss and water, which froze to produce a hard coating that allowed the sledge to glide over the snow.

There are considerably fewer dog teams in the Arctic now due to the convenience and greater speed of the skidoo, or snowmobile, and in many Arctic regions, teams are now only kept to entertain foreign visitors. However the skill in breeding and training dogs for a successful team is no less than it was in the past. The convenience of skidoos, owned by a large percentage of the Inuit population today, is countered by their tendency to break down and consume large amounts of expensive fuel. Many Inuit still bemoan the fact that a skidoo cannot be eaten if all else fails.

Navigation

In conditions that include twenty four hours of darkness, complete white outs, the inability of a magnetic compass to work so close to the North Pole and, until recently, no maps, the Inuit means of navigation had to be effective and dependable. These ranged from using the night sky during the dark winters, following the sun, following ice floes and the patterns that the wind had made in the snow. In the Pond Inlet area the prevailing wind is remarkably constant so by examining the build up of snow and ice around an obstacle they can judge direction extremely accurately. Folklore from other Arctic areas says that

The tailoring skills of Inuit women are legendary. Working from raw hides – caribou, seal and polar bear – they fashion clothing. Since the fur clothing is used in a cold, dry environment there is little need to tan the hides. Instead they scrape the hides to soften them, using a metal bladed scraper. Once softened, even the heaviest caribou skin can be worn comfortably close to the hunter's body.

The threads with which the clothing is sewn together come from the tendons that run along either side of the caribou's spine. These are the strongest, most durable fibres. Sinews are still used in preference to the strongest button thread available in the Pond Inlet store.

directions from point to point were passed down through the generations and fashioned into stories so that none of the details would be forgotten.

The inland lakes on Baffin Island contain delicious arctic char. The problem is how to fish for them when the surface of the frozen lake may be nearly two metres thick. The Inuit solution is to cut through the ice with ice chisels and picks. These tools are heavy and sharp, the chisel blade being five centimetres wide and the ice pick, a triangular sectioned needle pointed spike, two and a half centimetres thick. Using these, the ice is gradually chipped and broken, the chippings being cleared away until the breakthrough comes and water rushes up through the hole to the level of the lake surface. Having made such a hole, a specially designed board is dropped through the hole attached to two cords: one to be carried under the ice, the other to operate a lever on the board. This lever, progressively dropped and hoisted, forces the board, which floats up against the under surface of the ice, to creep along. As it creeps along, other fishermen listen for its presence. Once it has been sent a sufficient depth, it is located and retrieved by the digging of a second hole. With two holes and a rope threaded between them, it is now easy to lower and raise nets under the ice. Ever resourceful, these 'jigger' boards are now being fitted with radio locating devices.

The hides are carefully cut to shape using a semicircular bladed knife called an ulu. Patterns are carried in the head, collated from years of tailoring experience.

Pond Inlet

Pond Inlet is typical of modern Inuit towns. Over 660kms north of the Arctic Circle, it relies on supply runs by boat during the summer when the shipping lanes are ice-free and by plane in winter to provide it with all the basic equipment required to support a town with an expanding population. With over 1100 inhabitants, the majority of whom are Inuit, families are living in modern, heated houses and all local amenities are provided. Tourism is on the increase and provides a much needed income. Traditional practices are proudly still retained here and an official ranger service of local Inuits is on hand to provide emergency services and to train pilots in Arctic survival skills.

The Inuit have been prey to many Western pressures, in particular during the past few decades when governments have wanted to bring them into line with modern mainstream lives and thinking, believing them wrongly to be ignorant of the value of their natural resources. Most of the people of Baffin Island,

Digging through the Ice

Inland fish can be caught with netting: the problem is cutting through the ice. Stones at the lake edge mark places where fishermen have had good catches.

including Pond Inlet, only settled into permanent communities 30 years ago when government officials persuaded them to so that their children could attend school. Before that, most still lived in small family camps scattered across the region. By the end of the 1960s they were part of the Canadian welfare state and were the recipients of health care, education, modern housing and public facilities but, with little control over their own lives compared with the past – a confusing compromise between this and their indigenous way of life. Most people are now wage-employed, most by the Pond Inlet co-operative.

Inuit economies have been further effected by the major swing in the seal fur trade. Heavy demands for harp and hooded seal pup fur during the 1960s and 1980s meant slaughter was on a massive scale until the EC banned the trade in all sealskin products. The market subsequently collapsed and now sealskin is only used locally, determined by food needs.

Many of the families in Pond Inlet still go out in the summer to live in traditional tents on ancestral sites to hunt and fish, and all year hunting provides an important food source. But their dependence on wildlife is at the centre of a controversy between environmentalists and industrialists regarding the future use of

1 The surface snow is cleared away.

2 A hole is cut into the ice with a two metre, triangular pointed ice pick and a similarly-sized chisel-bladed ice chisel. The ice chippings must be removed every few minutes as the hole deepens.

3 The hole is cleaned up carefully, so that neither man nor the tools are lost. The water floods into the deep hole until it is level with the surface of the ice. Several fishing methods can be chosen. They are after the delicious arctic char.

Digging a Fishing Hole

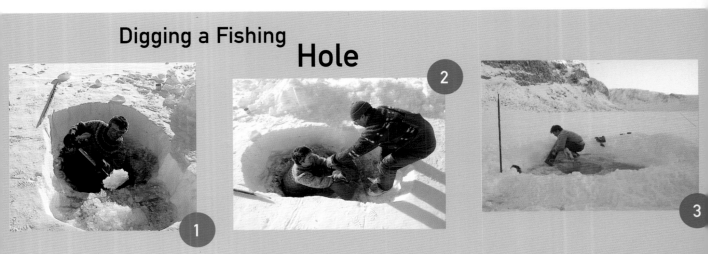

Solomon's Leister spear is fashioned from caribou antler and metal, lashed to a wooden shaft with a modern synthetic fibre. At the fishing hole the shaft will be lengthened by lashing another section on, so that the finished spear is over two metres long. At the ice hole, the spear is placed in the water. A small walrus ivory lure is dangled in the water to attract the fish to the spear. The fisherman aims to spear the fish on its back, the arms of the spear flex allowing the metal spike to impale the fish while the arms hold it securely.

Pond Inlet

Some **>** things never change: the chill Arctic sunset strikes the same notes in the human spirit now as it did when mankind first arrived in these latitudes.

> Today there is no human being on the planet that has not been influenced by the technological developments of recent years. For the Inuit the changes have been profound. Pragmatic by nature, they have been swift to embrace technology that makes life easier. Now they face the difficult task of striking a balance between their aspirations and their cultural heritage, which may seem outmoded and obsolete. Many of the younger generation are frustrated by the difficulty of funding their futures in such a remote community.

the Northwest Passage. There is pressure being exerted to ensure that Lancaster Sound, with its prolific wildlife resources, is kept open by ice-breakers for longer in the year to facilitate the extraction of ore and oil. The effects that this will have on the environment, wildlife and the Inuit way of life are unknown and feared. The young people in Pond Inlet, although receiving a modern education, are encouraged to learn some of the more traditional skills. However, they insist that they are no different to any other young people. 'It's hard to keep our traditional values mixed in with what we are today'; 'We don't travel everywhere by dog team or live in igloos'.

An ancient people, the Inuit have lived in the Arctic successfully for thousands of years and perfected a life in the cold in balance with natural resources and according to tradition. Being an Inuit

'means being able to understand and live with this world in a very special way. It means living with the land, with the animals, with the birds and fish, as though they were your sisters and brothers. It means saying the land is an old friend and an old friend your father knew, your grandfather knew, indeed your people always have known … we see our land as much, much more than the white man sees it. To the Indian people our land really is our life'.
Richard Nerysoo, Fort McPherson 1977

In a land of snow nothing lives forever, even Jacob's ice sculpture is now gone save for this image. For us it may be hard to grasp the impermanence of this land but for the Inuit it is a reality of life they accept, live within, and are themselves shaped by.

Arnhem

Entering Arnhem Land is like stepping back in time: Australian Aborigines maintain the fundamentals of a lifestyle that has not changed for 60,000 years. The simplicity of their approach to life is inspirational.

Land

Arnhem Land

The word Aborigine comes from the Latin 'from the beginning', aborigin. It is a name used for people all over the world who were there first, but especially for the native people of Australia. The term was first coined by the Europeans who arrived 200 years ago, but Aborigines believe they have been there from the beginning of time, known as The Dreaming, when their ancestors first walked across the land.

Mangrove swamps and flood plains form the northern coastline of Arnhem land, a potentially very rich area for 'bushtucker' if you know where to look.

Further inland the flood plains abut the stone country. Here in the rainy season, known locally as 'The Wet', Aborigines took refuge under cliff overhangs.

Human survival on the Australian continent is inextricably linked with the Aborigines. Their inherent ability to live by and with their natural resources is well documented, how they interact with the environment and shape it to suit their needs but never destroy it in the process. Their antiquity is acknowledged, as there is evidence to show that people have been living on the Australian continent for between 40 and 60,000 years and this great history and their beliefs, are minutely described in the thousands of rock paintings found throughout Australia – artwork that is well recognised by the art world and now highly sought after. In terms of safeguarding the resources available to them, the Aboriginal lifestyle has been recognised as one of the world's most faultless methods of conservation.

Aboriginal History

The word aborigine means 'the ones who were there from the beginning'. It is the accepted name for indigenous people worldwide, although it has become synonymous with the native people of Australia. The Aborigines believe they have been in existence since the beginning. They speak of the creation as 'The Dreaming', a sacred time when the Ancestors walked across the land and made all its elements, the sky, the land, the plants, the animals and its people. Before the Ancestors came into the world

These stick figures painted in red ochre may be several thousand years old. Recent archælogical evidence suggests that the Aboriginal settlement may have begun over 60,000 years ago.

it was just a flat emptiness. Every natural feature – a rainbow, the sun or the moon – has a meaning and a story behind its presence. After creating the world, the Ancestors, who could take on animal and human form, sank back into the earth from where they came. Although the period of the creation is over, 'The Dreaming' is eternal and the Ancestors are still alive as their bodies remain in the rocks and trees from where they give power and life.

European Contact

Arnhem Land was so-named by the Dutch explorers who were the first Europeans to land in Australia (which they called New Holland). In 1605, a Dutch ship sailed into the Gulf of Carpentaria and Arnhem Land was visited again in 1644 during one of the celebrated explorer Tasman's Pacific expeditions. At that time it was widely believed that there was an unknown continent in the high southern latitudes, and it was not until Cook's first voyage that Europe would learn that Australia was an island. There was to be no European settlement of Australia for over a century, although a succession of Dutch East Indiamen came to grief on the coast of western Australia and evidence from the wreck of the *Batavia* shows that survivors camped ashore for some time. Tasman recorded 'nothing profitable' anywhere in New Holland, so it is perhaps no surprise that further exploration was a gradual process. British and French interest in the region increased in the mid-18th century. In April 1770 Captain James Cook landed at Botany Bay in east Australia.

Britain's commercial, strategic and social needs inspired the 'First Fleet' which founded the Botany Bay penal settlement in 1788, landing just six days before a French expedition reached the same area. Botany Bay expanded in grim circumstances: shipping thousands of convicts to the colony suited the needs of the British criminal justice system, but the human consequences were appalling. Relations between the aboriginal people and the settlers varied

Aboriginal art represents not only the animals that were hunted but also spirit beings, for Aboriginal life is held in an intricate balance with the forces of nature.

How to trap Fish

The Arnhem land streams abound with fish. The region experiences very high tides which flood many miles inland. Taking advantage of a tidal change, Aborigines from the outstation of Yilan set a traditional fish trap.

from mutual suspicion to official acts of genocide by the colonial authorities. Tales of aboriginal cruelty were played up by the latter, as it obviously discouraged convicts from escaping into the bush. In the 19th century there was an influx of settlers, and the policy of transportation for criminals came to an end. The new settlers were mainly sheep famers who shipped the wool back to the English wool mills. They were then followed by the cattle and arable farmers, who claimed vast tracts of land as their own. In the 1850s gold was discovered in New South Wales which heralded the initial gold rush.

With this influx of white settlers, many laying claim to land that had up until that time been solely used by Aborigines, conflict soon arose. The conflict was rooted in the concept of ownership. The Aborigines, knowing that the land had been created by their ancestors, believed it was theirs, whilst newly created Australian laws gave the Aborigines no land rights at all. Large areas, or Aboriginal Reserves, were eventually set aside by the government for use only by Aborigines, though much of this land was 'given' to them only because it was considered useless and unprofitable. There was no regard for the Aboriginal ancestral lands and their sacred places.

From an estimated population of 300,000 Aborigines at the time of the arrival of the white man, the population dropped to 60,000 by the turn of the 20th century. The loss of land, loss of lifestyle and western disease accounted for a large percentage of this decline. In 1979 the estimated population was 150,000.

By the 1970s the plight of the Aborigines had reached the world stage and protestations by human rights groups began in earnest. In 1976 a referendum gave the federal government the power to pass laws on Aboriginal affairs in all states and Aborigines were included in the census. At the end of the decade some Aboriginal reserves, including Arnhem Land, were declared by law to be Aboriginal Land. In Northern Territory the Aborigines gained title over an excess of 100,000 miles^2.

Arnhem Land

Australia covers a wide range of environments, from tropical coast, rainforest, and woodland to open scrub and desert, and the Aborigines once thrived in all of them. Their skills at finding food

The river is dammed with stakes and closely-woven matting. On one side of the matting a basket is set. When the tide changes, the current will sweep towards the basket entrance and fish carried with the current will become trapped in the basket.

Spearing Fish

Waiting almost motionless, Michael is poised to spear a barramundi (Giant perch) *Lates calcarifer*, using a fishing spear and woomera (spear-thrower). The fishing spear is in constant use for catching fish, crayfish and long-necked turtles. At night whole communities spear fish by torchlight.

through hunting and gathering were highly adapted to suit each individual environment and its seasonal variations. As white settlement disrupted traditional Aboriginal life in Australia in most areas, hunting and gathering is now largely confined to the more remote areas of northern and central Australia. Western life has of course had a major impact in these areas too, in particular providing packaged foods so that there is no longer a reliance on foods collected from the bush. But still, large amounts of bushtucker are collected and preferred, some by methods that are little changed from the past or are just assisted by the use of a vehicle, a motorboat or firearms.

Arnhem Land, in the extreme north of Northern Territory, is one area where many of the traditional values and practices of the Aboriginal people remain intact. It has been subjected to change, but in the late 1970s a large percentage of the Aboriginal people living a modern existence in and around the town of Maningrida in the north east, moved back to their bush stations. This was a move back to the traditional lifestyle and away from the problems created by bringing a semi-nomadic people into a sedentary lifestyle, a life with social security payouts but no purpose. The name Maningrida is the anglicised version of the Ndjebannaname Manayingkarirra which has the meaning of 'The place where the dreaming changed shape'.

Arnhem Land has been protected by its remoteness and isolation. The tropical environment was harsh and hostile and several early white men died of malaria within days of arrival. Matthew Flinders was the first arrival in 1814 and although the coastline was officially charted in 1827 by Captain Phillip Parker King, no actual settlement was established. Three unsuccessful attempts at permanent settlement were made in the west of Arnhem Land between 1824 and 1849. Ideally the British wanted to create a trading post and a strong northern settlement to protect them from the Dutch. By the 1920s the missionaries had arrived, but the impact on the indigenous

Johnny fashions a new fishing spear. The shaft is a macaranga sapling; the four prongs are made from any available metal rods. After stripping off the bark and straightening over the fire, a hole is burned eight centimetres into the thick end of the sapling with a hot metal rod. Into this the metal prongs are fitted and bound. To tighten the arrangement and splay the prongs, a wooden wedge is driven between them.

Arnhem land is a timeless place: while Johnny's spear has metal barbs and fishing line binding, it is modelled on an ancient pattern which had ironwood prongs and was bound with banyan bark fibre and beeswax.

population was small. In the 1930s Sid Kyle-Litttle established the first trading post after an epic and dangerous journey to reach what is now known as Maningrida, and a town began to develop. By the early 1970s the settlement that originally was only intended to provide trading and medical services for the area, was supporting a population of 1200 people of which only 120 were European.

In 1972 the Labour Party gained power and Aboriginal rights were high on their agenda. The Aborigines realised that unless they returned to their lands they could lose them. This was coupled by a general disenchantment with their way of life, the quarrels and unhappiness that an urban existence had brought, a far cry from the pre-European way. Many therefore returned to their own territories, some only seasonally, to re-establish the outstations and their previous lives, gathering from the land and living close to the sacred sites that are still of great importance to them.

Aboriginal Culture

Individuals belong to family groups or clans and are divided into two sides, or moieties. Marriage can only take place between opposite moieties. Not only people, but things are also given a moiety status. In north east Arnhem Land the Aboriginal people belong either to the Dhuwa or the Yirritja moiety. Clans

How to cook Goanna

are often called after an animal, either because that is the shape their ancestor took on, or because they believe the ancestors had a strong connection to that particular creature. That animal is then afforded a totemic status. In northern Arnhem Land they believe that the spirit of an unborn child will appear to the father whilst he is hunting in the form of a particular animal. After the birth of the child he or she will have kinship with that animal. Boys would come to the end of their childhood in their mid-teens at which time they would be taken through their initiation ceremony, the details of which are never divulged. Marriage would have been an arranged affair at around 25 years of age, with the woman living with the man's family beforehand for a year or so.

Every clan has a place that is sacred to them, a place where the Ancestors lived and performed their ceremonies. Ceremonies would bring people together from far afield and may last for several days as people share, exchange and distribute food. The most important ceremonies are those that involve the adult men asking for the power of the Ancestors so that all of life, the people, plants and animals, will develop as they should. Women too are afforded their own exclusive ceremonies. It is the role of the elders to arrange the great rituals and to pass on the traditions. The elders are assumed to be wise and all the young men are expected to heed their word. Traditional songs play an important part in life too. These may be sacred and inspired by the Ancestors or made up by an individual at the time. Sacred songs, stories and beliefs frequently appear in their paintings and are the basis for Aboriginal art.

Aborigines had few possessions and offences for which punishment was required were seldom. The harshest

▲ **Like the majority of the Aboriginal diet goanna is cooked by simply casting the animal onto a hot fire and roasting it in its skin, turning it occasionally by the tail until it is ready. Once cooked, it is butchered in an organised way and the white tasty meat is shared out. It may seem unhygienic, but in fact it is a sound way to prepare food: there is the absolute minimum of food handling, it is cooked at a high temperature and eaten straight away.**

▲ **To make a good bed of embers, the fire is loaded with large lumps of magnetic termite mound. These will burn like charcoal and impart a slight flavour to the food.**

punishment would have been derision and ridicule from others which would have been inflicted by family members.

A death will be accompanied by much wailing, mourning and self-wounding. Aborigines believe that a part of the person survives and this part will return to the place where they came from. In Arnhem Land still, part of the body, possibly bones, will be kept for months or even years after the death, maintaining the person's spirit close by. Here, the person will go to the Land of the Dead which is across the sea.

The Aboriginal Way of Life

Much of Arnhem Land is covered in open-canopy forest but it also supports a wide range of habitats from coastal mangrove swamps, saltwater estuaries and dunes, to rainforest and open plains. It consequently offers a wide variety of species of plants and animals that provide an essential source of carbohydrate and protein to the current Aboriginal population and in the past, materials to make fire, shelter and tools and provide water.

The outstation of Jibalbal is 40kms from Maningrida, its terrain includes the west bank of the Blyth river which is an excellent source of fish and the mouth is a sacred area of Mermaid Dreaming. The traditional Aboriginal way of life here ensured that they ate a rich and balanced diet of seasonal fruits, nuts, roots, vegetables, meat and fish. Land and freshwater

Michael is making fire by friction using a hand drill. He learnt this technique when he was a teenager. At that time, this was the normal way to make fire. Today, younger generations look on with interest. They are more used to using matches or lighters.

The women are making digging sticks from ironwood (as tough as its name suggests). Today, metal digging sticks are the norm, but when needs must, the old skills are employed. In times of war, the digging stick was the woman's weapon.

Surviving the Mosquitoes

animals and birds hunted in this part of north east Arnhem Land included wallabies, buffalo, goannas, geese, ducks, emus, file snakes, bats, flying foxes and long-necked turtles. Marine and estuarine animals included fish, dugongs, dolphins and whales, bivalves, oysters and crabs and insect foods such as honey (also known as 'sugar bag'), moths and grubs.

Whereas Aboriginal groups are now relatively settled and living in permanent communities, patterns of the past involved the movement of clans that were dictated by the seasons. As food sources became scarce in one area and abundant in others, people moved to exploit them. The area utilised would depend on the richness of the resources, with those living in the poorer central arid regions needing to cover several hundred square kilometres to survive. This seasonal understanding of the environment is as much key today as it was in the past. The introduction of modern technology may have increased the efficiency of hunting as shotguns replace spears, fishing nets replace bone hooks and fibre-string nets and vehicles allow much greater distances to be covered, but this is still no substitute for an indepth knowledge of your natural resources when searching for food.

⚈ The Arnhem land forests are typical of open tropical woodlands: the mosquitoes are terrible. To prevent them biting, Aborigines learned to make smudge fires from termite mounds near to their shelters.

Aborgines recognise more than the two standard seasons of wet and dry in Arnhem Land. There are a number of phases or seasons each of which is an intricate balance of the weather pattern and the availability of plants and animals. Changes in their surroundings would indicate the optional times to crop sources of bushtucker and the Aborigines would watch carefully for these seasonal indicators; in particular the flowering of plants. When the orange blossom of the batwing coral tree falls, the women know it is the right time to dig crabs under the mangrove mud; the appearance of milky white 'oyster flowers' tell the people that it is time to move to the oyster beds for their seasonal collection. When certain wattles are in flower it means that the sea turtle and stingray are fat and when some eucalypt trees are in flower this indicates that honey is ready for collection.

Despite changes in method, the roles of men and women in the provision of food are still clearly defined. Men remain the providers of land and sea mammals and are fishermen whilst the women gather plants, shellfish and hunt for small mammals. The women do provide the bulk of the food, mainly in the form of carbohydrate, more than 60% of the total consumed.

Aboriginal spiritual life is closely interwoven with food, and customs guide many aspects of the process of gathering, cooking and eating. Young people are taught that it is taboo to forage in certain areas, and animals in these areas are not to be disturbed. Other areas can only be used when hunting for ceremonies and others are only open to initiated men. The totemic system usually forbids people from eating their totems except during certain times of the year or at particular ceremonies. In east Arnhem Land each family group had only one totem which was inherited from the father, people were allowed to eat their totemic species, but only in moderation. Each group would caretake for its totemic species and during ceremonial practices would encourage the species to multiply.

Tools

The Aborigines needed few possessions, as their weapons and implements were so ingeniously designed that only a few light tools would be necessary to equip them with the means of obtaining all their daily requirements; there was little to incumber them in their nomadic lifestyle. Often one implement, such as a throwing stick, would double as a digging stick or even

This termite mound has been broken open to obtain the darker matrix inside it. This will be burned on a small smudge fire beneath sleeping platforms to deter mosquitoes. Additionally the mound itself will be set smouldering.

Mangrove Worms

The mangrove forests abound with life: everything seems to be moving. The most unlikely source of food is found inside the trunks of fallen mangrove trees.

as one half of a fire saw. Beautifully decorated Aboriginal weapons and utensils are found in many museums today, the intricate designs on which tell of the clan's stories and their 'Dreaming' sites; craftsmenship that went beyond practical requirement.

In Arnhem Land, boomerangs were not utilised, instead a spear and spear-thrower, or woomera, were used. The woomera fits on to the butt of the spear and works like an arm extension making the throw more powerful. Spears were used throughout Australia for personal defence, fighting and punishment as well as for fishing and hunting marine and land animals. Depending on the need, these were made in many different designs with either simple fire-hardened tips, spearheads or multiple-pronged, using hardwood, hide, flint or glass heads attached with gum. In Arnhem Land a four-pronged fishing spear is made, traditionally the spikes would have been made of ironwood, lashed together with fibres and secured with beeswax. Throwing sticks were also used, which knock out the prey animal rather than kill it. This would fly in a straight line, unlike the boomerang.

The women's main tool was the digging stick which is still used for a multitude of purposes, not least for digging out root

This is the teredo or mangrove worm. An Aboriginal delicacy, it is eaten raw, just as you see it here. It tastes like crab pate with a hint of wood.

vegetables such as yams or crabs in the mangrove forests. Metal crow bars have replaced the more traditional design, though the methods of foraging remain the same.

Fire

Unlike in the Western world, Aboriginal children have always been encouraged to learn to use fire in play. They may now be seen using a box of matches or lighter to light a bush fire, but in the past fire was started using a fire saw or bowdrill. Fire plays an essential part in Aboriginal life as firing the landscape is used to aid hunting and encourage the growth of plants. Areas with different burn age would provide a variety of plant foods and animals could be hunted when cropping the new shoots. Firebreaks would be used to hunt kangaroo or wallaby for

ceremonies during the dry season. A large group of people would hunt together, some would light the fires whilst others waited in ambush downwind – an effective way of killing a large number of animals, generally now replaced by using vehicle spotlights at night.

Water

Even during the dry season, river water is still available in north east Arnhem Land. However, the rivers are inhabited by crocodiles and there is a saying that if you use a particular path to the river on more than one occasion, one of them is sure to get

The mangrove worm is not a worm at all: in fact, it is a mollusc. Its shell is modified to provide the equipment it needs to bore through timber. In the 18th century, these molluscs were the bane of mariners.

Foraging for Food

you. Water can also be collected from the paper bark tree where it collects in a bulbous swelling under the bark of the trunk. Using an axe to break into the swelling, pure, filtered water will run out.

Shelter

The Aborigines may now spend the dry season only on an outstation and during that time they live in permanent dwellings. In the past a dry weather shelter was constructed using a wooden framework that supported a platform which was then protected by a roof made of bark from the paperbark tree. Arnhem Land mosquitoes are large and numerous and protection from them was gained through burning pieces of termite mound underneath the platform which gives off an effective smoke screen. Shelters would have been built to house all members of the clan with boys as young as seven years old having a shelter of their own. Wet season shelters would have been a great deal more sturdy than their dry season counterparts, circular and squat they would have been completely covered with bark.

▽ **Looking out at the open forests of Arnhem Land from a stone country outcrop, it seems hard to imagine that it is possible to find food in the surrounding countryside.**

The Women and Their Work

Although women do now grow their own vegetables on the outstations as well as using shop-bought carbohydrate, the food provided by bushtucker is still essential to their dietary needs. Gathering work is extremely intensive work, using their digging sticks they forage,, collecting their bounty in dilly bags which are made from the bark of the aerial roots of the banyan tree.

As well as using fire to encourage regrowth, some Aboriginal women when collecting yams will always leave a small piece of the yam in the ground. This is based on the belief that, if you remove the whole yam you will anger the food spirits and they will not let any more yams grow in its place. Yams are an important source of carhohydrate and they are collected during the dry season. A few dry tendrils or withered heart-shaped leaves above ground are the only indication of the food source held in the soil below. There are two types found in Arnhem Land, one is eaten raw, the second type needs to be washed, cooked, sliced and then re-washed to remove the poisons it contains before it can be eaten.

Before the availability of flour, a paste made from ground seeds would have been baked into unleavened bread or 'dampers'. The type of seeds used varied across Australia, many types requiring much preparation to make them suitable for eating.

Isabel has made a hole in this snail shell to turn it into a specialised yam grater.

Like children everywhere, the ingredients often get nibbled before they are cooked. Here, they are munching on long yams that taste similar to potato.

Supporting a steam-softened cheeky yam (poisonous) she grates it finely with the snail shell. The gratings will be soaked overnight to remove the toxins.

Making a Dilly bag

🔺 **Stretching the weave across her legs, Isabel is making a Dilly bag (food gathering bag) out of banyan fibres. These bags are now more likely to be found adorning the walls of a trendy New York apartment, than slung over an Aboriginal's shoulder.**

In the open woodlands and rainforests the greatest number of tubers, fruits and nuts are found, including figs, wild gingers, wild plums and palms. The wild Arnhem Land plum has been found to contain very high amounts of vitamin C. Towards the end of the wet season, women collect waterlillies from the billabongs on the open flood plains, for their seeds, stems and corms. Also found here are the long necked tortoises which hibernate during the dry season and are located by finding indicative marks on the ground surface; they are then dug out of the ground.

In addition to crabs sought after in the mangrove swamps, the mangrove trees themselves are a rich source of nutrition in the form of mangrove worms. These large white slug-like animals are in fact crustaceans. The shell-like head is not edible but the remainder is eaten eagerly. Also found in the mangrove trees are oysters and bivalve mussels.

Foraging groups tend to be single sex groups of relatives and friends. Men would occasionally hunt alone and sometimes with a wife and their children. The young children would usually accompany the women and boys older than ten go with the men. Large groups would be formed to take advantage of particular resources such as wallabies and goannas captured with firebreaks and for ceremonies. Goannas would also be scented-out by dogs, caught with throwing sticks and more likely today, shot. Men and women would both hunt for file snakes. Snakes that live in the water and are so named due to their sharp scales that will literally file away a hunter's skin; traditionally they were caught by allowing one to wrap itself around an arm and then quickly lifting the arm out of the water.

Hunting and Fishing

The hunting of large mammals, birds and fishing is the male domain. Firearms are frequently used now, as are vehicles, to travel to the most productive hunting grounds, but spears and throwing sticks are still made and used. As the people preferred a varied diet, the hunters would initially seek a different species

Raffety Fynn, who provided the film crew's logistical support, takes a quick drink from this paperbark tree. The bulge in the trunk indicates that, if cut into, this tree is likely to hold safe drinking water.

Steaming Yams

Long yams and cheeky yams are prepared for cooking on a fire of hot embers.

Covered with paperbark, they are sealed over with earth and left to steam for 25 minutes.

After steaming they are uncovered. The paperbark has kept the dirt off the yams. The long yams can now be eaten and the cheeky yams progress to another stage.

from day to day. Dogs would have been used and a sign language was adopted during hunting when tracking and then stalking an animal.

Wallabies on plains are herded by burning, and are mostly shot. Macropods were typically hunted early in the morning or late in the afternoon when they would be feeding on the flood plains. The usual method was to stalk the animal, sometimes using dogs to scent and give chase before the prey is shot or speared. Macropods would be hunted using firebreaks when providing food for ceremonies.

Ducks, turkeys, cassowaries and other fowl are hunted in the grasslands and open canopy forests. Magpie geese were hunted with throwing sticks. Hunters rise early in the morning so that they can hide in trees whilst it is still dark. As the magpie geese took off at first light the hunters would try to hit them with throwing sticks. The procedure now is similar, but the geese are shot instead. Emus are extremely inquisitive birds and a number of inventive ways have been devised to catch them. At Jibalbal the people tend to shake the branches of trees to attract them.

Saltwater and freshwater fish are an important part of the Arnhem Land diet. Many species migrate up the freshwater reaches of the rivers and are speared in the billabongs. Barramundi are such migrants and are hunted by spear as the waters recede. Hunting would be carried out alone, in small groups or more usually in a large group. Traditional boundaries would determine the areas used, though often these areas would be reciprocally shared. Generally they would take as many barramundi as they could, rather than leave them to the birds as the creeks receded. All those partaking in the hunt would be given their fair share of the kill.

The Arnhem Land coast is a traditional hunting ground for dugongs and these and turtle are still hunted using traditional harpoons, though motor boats have replaced dugout canoes. Buffalo, introduced in the 19th century, and feral pigs provide an important protein source. They are both now hunted with shotguns.

Cooking Methods

The most common cooking methods are roasting on coals, cooking in ashes and steaming in a ground oven. Roasting on hot coals is used to cook smaller animals and as an initial stage in cooking larger animals. The animal would be thrown onto a fast-burning fire which within ten minutes will singe the fur. The animal would then be taken off the fire, the intestines removed and the remaining fur scraped off with a sharp implement. The animal is then returned to the bed of hot coals to be cooked.

The most favoured method of cooking large game, including whole kangaroo, pieces of buffalo and emu, is in a steam oven. In a large pit a fire is kindled and allowed to burn down to a bed of glowing coals, large lumps of termite mound are then added to the coals. The meat is placed on top of this, then covered by layers of branches and paperbark and the whole is left to steam.

The Future

A better understanding of the Aboriginal way of life and its many mystiques is slowly being unravelled through research and working much closer with its people. Efforts are being made not only to fully integrate them into the 20th century Australian way of life, but also to allow them the freedom to maintain their traditional practices and values. But this is not without its problems and contradictions.

In north east Arnhem Land where traditional practices are still in evidence and respected, the Aboriginal population own their land and have the authority through the Bawinanga Aboriginal Corporation to decide what changes can or cannot be made, in order to promote Aboriginal culture and sustainable enterprises. The Aboriginal way of life has reverted back to the outstations and they are proud to maintain this traditional lifestyle save for the addition of a few modern artefacts.

The whole village took great delight watching the rushes of the day's filming. Aborigines have a sharp sense of humour and mercilessly poked fun at each other's performance.

In survival terms the Aborigines, without the white man's interference could no doubt have thrived for another 30,000 years and still not made a serious impact on their surroundings. Such a profound and deep ecological knowledge should not now be isolated and forgotten. There is a great deal to be learned from them, not only in terms of conservation but in their spiritual values and deep respect for the land created by their Ancestors.

'Our life is like a circle, we follow the seasonal cycle. Each season has it's own way of telling us the best place and time to hunt. Everything has a role telling us when it is the season to hunt or not to hunt a particular species ...'

Siberia

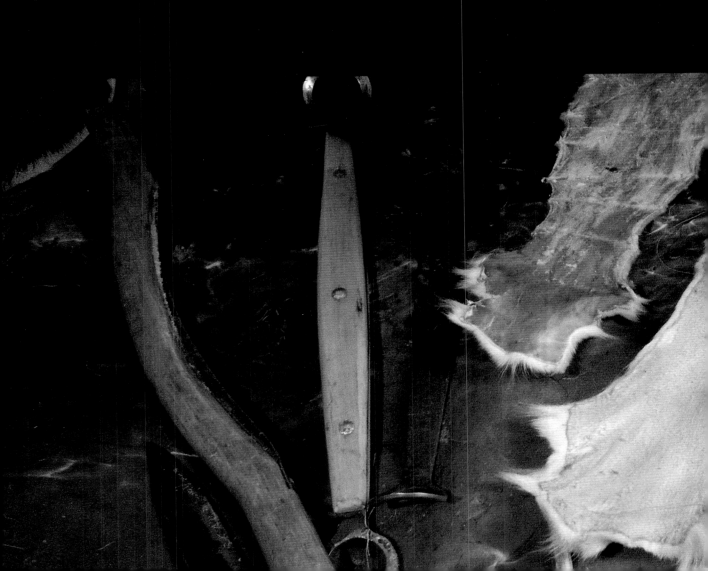

Siberia is one of the coldest places inhabited by human beings. Life here is made possible by one thing, the reindeer: the food it provides and the knowledge of how to make warm clothing from its fur.

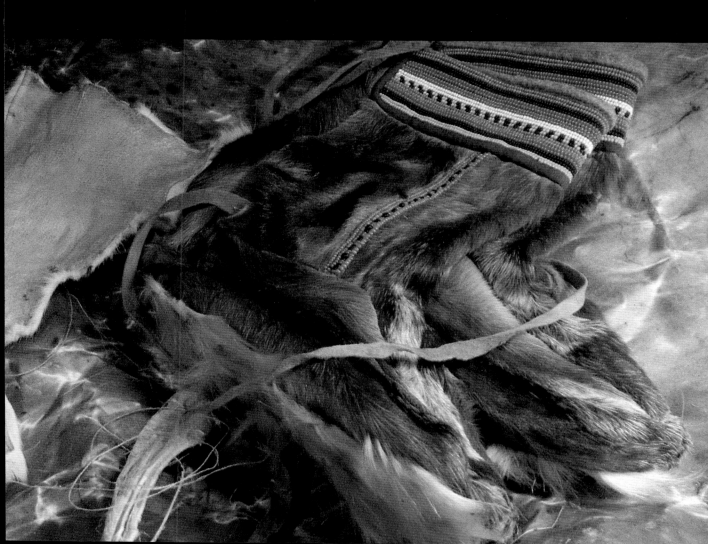

With a flash of colour, the needles of the larch trees turn yellow heralding the onset of the feared Siberian winter. The Evenk, who inhabit this vast region, are gentle reindeer herders. Constantly moving their reindeer to fresh lichen grazing, they are some of the last nomads on earth that live in tepee-style tents. Nomads to the the core, they say that the best thing in life is arriving at a new place.

⬆ **A tent with a wood-burning stove and an axe are the vital tools that make it possible for the Evenk to endure the harsh cold of Siberia.**

The origin of the name 'Siberia' is said to have come from the word *sibir* which in the Mongolian Altay language means 'sleeping land'. Mention the name Siberia and it still evokes powerful images predominantly of vast, freezing cold tundras and hard-working Russian exiles. Now synonymous with the Transiberian Railway and gas and oil extraction on a huge scale, Siberia is still home to the few surviving nomadic brigades of Evenk people, one of a number of indigenous groups who originally existed in the immense expanses of taiga forest in the Siberian north.

The Place

Siberia is a country of huge dimensions. With an area of over five million square miles, it is one and a half times the size of Europe and forty times the size of the UK. Now a part of the Russian Federation, Siberia is more of a geo-political invention than a physical geographical unit. It includes not only a broad spectrum of landscapes but is also home to people from very different cultures.

Geographically speaking, Siberia is made up of part of one of the oldest landblocks, Angaraland, and a section of the shore of

◀ **Siberia is an enormous wilderness: here in the Evenkia region there are thousands of lakes, millions of larch trees, but very few people.**

The most fundamental focus of Evenk life is the reindeer. Until a thousand years ago the Evenk hunted wild reindeer; today they herd them for their meat and pelts, harness them as pack animals, and even ween their infants on reindeer milk. For an Evenk, life without a reindeer is inconceivable.

Skinning and Butchering

▲ **Gentle folk, the Evenk are some of the last people living full time in tepees. When told that many people in Europe envy them their lifestyle they said, ' They should be here in the winter.'**

the Arctic Basin. It slopes gently up to the interior plateaus and all the rivers flow northwards. Russia is located to the west, the Arctic Sea to the north and to the east Japan and Korea. Its main mountain ranges are the inaccessible Altai in the south, Western and Eastern Sayan stretching from Kazakhstan to Lake Baikal and a number of ranges from the Pacific Coast to the Maritime Territory. It spans eight time zones.

The journalist George Kennan wrote in his book *Siberia and the Exile System* (1891),

> 'you can take the whole of the United States … and set it down in the middle of Siberia, without touching anywhere the boundaries of the latter's territory; you can then take Alaska and all the countries of Europe, with the exception of Russia, and fit them into the remaining margin like the pieces of a dissected map. After having thus accommodated all [this] you will still have more than 300,000 miles^2 of Siberian territory to spare … an area half as large again as the Empire of Germany.'

Like all ▶ inhabitants of the northlands, the Evenks are skilled butchers: nothing goes to waste and the hides are carefully removed so that they can be turned into some of the warmest clothing made anywhere in the world.

Little is ▶ wasted in the butchering. For example, the liver and kidneys of a reindeer are sliced through and hung on a nearby branch to dry, well out of reach of the camp dogs.

▲ Anthropologist Thomas Johansson explains why there are very few people in Siberia: 'the search for calories is so difficult. This leg of reindeer meat would enable four people to live for one day, whereas the 6kg of edible lichen beside him would feed only one person for one day.'

Some of the largest rivers in the world cross Siberia, the Ob, Yenisey and Lena, are three of the 53,000 rivers in total. It has more than one million lakes including Lake Baikal, the largest freshwater lake in the world; 420 miles long and 90 miles wide it covers 12,441 miles^2 with an average depth of 5404ft.

Taiga, the vast dense forests of birch, pine, spruce, and larch, cover huge areas of central and northern Siberia. To the south are the flat, dry steppes that reach into Central Asia and Mongolia and to the far north large expanses of tundra, much of which is permanently frozen bog or permafrost.

Climatically speaking, the Siberian climate is truly continental with intense winter cold and great extremes of temperature between the seasons. The average January temperatures are -20°C to -25°C, dropping to -35°C. In July and August temperatures average 15°C to 20°C and can rise to 30°C. The first frost is felt in October with most snow falling in November and December. Annual precipitation is between 25 and 40cms with snow cover reaching between 50 and 80cms. North central Siberia is the coldest land in the world, with temperatures falling to -65°C, even colder than the Poles. With short summers and long freezing winters, Siberia experiences practically no autumn or spring.

Lean-to Shelter

➤ **If you found yourself stranded in the taiga, you could sleep next to a fire, but in bad weather you will need more shelter. One of the most efficient of these is the open fronted lean-to. Here is the bed and basic framework before thatching. Notice that there is no need for string or nails and that the bed is raised above the cold ground to a height of about 50–60cm.**

Stone Age Settlement

The first known Siberians were Palaeolithic, early Stone Age tribes who lived around Lake Baikal and the headwaters of the Ob and the Yenisey rivers. Evidence of Neolithic people, the late Stone Age, have been found all over Siberia. When the earliest Russians first encountered the northern people they were still living a Neolithic life. As late as the Iron Age, the steppes and forests from the Urals to Baikal were populated by tribes of Caucasian herders. Post-glacial remains have been found around the industrial town of Krasnoyarsk in central Siberia and these include mammoth and human bones, as well as weapons and ornaments.

By the third century BC, the south of Siberia was under the control of the Huns. Descendants of these people then moved west to terrorise Russia and Europe. In the first centuries AD, Turkic tribes moved in from Central Asia.

Russian Traders

The first Russians in Siberia were fur traders from Novgorod who reached the northern Ob river by the late 11th century. From the mid 16th century Russian influence began to take affect, following the success of Ivan the Terrible who claimed the Volga for Russia. The door to Siberia was opened. The following century saw an explosive territorial expansion fuelled by a lust for furs, quenched by Siberia's abundance of fur-

➤ **The shelter is thatched at the rear and the sides and bedding laid for comfort. At a distance of one full pace in front of it, a long log fire is built that supplies warmth to the whole length of the body. A shelter correctly built in this way enables rest and sleep even without a sleeping bag.**

bearing animals. People came in large numbers to the Siberian wastes and by 1639 they had reached the Pacific Coast at the Sea of Okhotsk; by 1697, the explorer Atlasov claimed the far north-east of Siberia for Peter the Great. The indigenous tribes may have found the newcomers a welcome change to the warring tribes of the Tatars initially, but with only bows and arrows against the Russian muskets they would have had no choice.

Russian settlement in Siberia was encouraged with promises of easy land and freedom from serfdom. From about 1650, the authorities began sending criminals to Siberia. In the 18th century, as Siberia's vast natural mineral wealth became apparent, those sent there were put to work to dig up whatever they could. Siberia became the grim destination for anyone who dared to challenge the Tsars. As the demand for labour increased during the nineteenth century, the list of punishable offences for which Siberia was the sentence, was increased to include prostitution, vagrancy and fortune-telling. The death penalty in Russia was then abolished and more people were sent to Siberia without a trial. At the beginning of the twentieth century, over 3400 exiles were being sentenced to hard labour weekly. Lenin himself spent three years in Siberia.

The greatest impact on Siberian development was the building of the TransSiberian Railway. Tsar Alexander III finally authorised the building of the 7500km line from Chelyabinsk to Vladivostok. Thirty years later it was completed and it still remains one of the world's most incredible engineering feats. Cities mushroomed along its length.

Lenin and Stalin expanded the Tsars' Siberian exile system into the world's largest concentration camp network, with resettlement programmes, transit prisons, labour colonies, and special psychiatric hospitals. The first camps were created to dispose of opponents of the Bolshevik regime during the civil war and were expanded by Stalin during the late 1920s. The system became known as the Gulag (from the Russian acronym for 'Main Administration of Camps': then a branch of the NKVD). Soon, a huge slice of north east Siberia was set aside exclusively for labour camps. The numbers grew during the

Birch trees grow particularly well in the northern forests. Birch bark is very useful, particularly for making containers for storing or carrying food. Arkadi is carefully peeling the outer bark from a birch tree, a job normally undertaken in the warmer months of the year, when the sap is flowing and the bark more loose. He is having to take great care not to split the bark sheet.

Making Footwear

1930s when Stalin successively purged the communist party and the Red Army. It is estimated that over 15 million people died in the Gulag, including over a million German and Japanese prisoners of war during and after World War II. The Gulag system continued well into the 1980s and how and when it was phased out remains a mystery.

The Indigenous People

The indigenous population of Siberia were nearly all nomadic herders. Those in the south herded sheep and horses while those in the north herded reindeer. Some groups were predominantly fisherman and hunters, hunting game inland and walrus, seal and whale on the north coast. Little agriculture was possible due to the prevalence of permafrost. Originally most would have practised shamanism. The Evenk people are one of 20 indigenous groups that followed this true nomadic lifestyle, now upheld by very few.

Russian interest in its indigenous people waned in the 19th century as the fur supply dwindled. Initial interest had been high and national homelands were created for the various groups which in effect saved them from being enveloped totally by the wave of immigrants. The larger groups, such as the Buryat,

The skin on the reindeers' shins is particularly tough and is therefore used for the manufacture of footwear. It is skinned with great care: scoring caused by clumsy cutting would weaken it.

Once the shin has been stretched with small sticks and hung in the chum to dry, it is scraped to remove a tough papery sheath on the inside of the skin.

The skin is tanned by rubbing the powdered wood of an old decaying larch tree (mixed with a little water) into the skin. The skins will be left rolled up like this for one or two days, before a final curing by smoke in the apex of a chum.

Yakut and Tuvinian, were granted their own Autonomous Soviet Socialist Republics which they themselves represented in Moscow. None of them had written languages until after the revolution.

The Evenk People

The Evenk or Tungus, which is the Russian name, are one of the most ancient of the central and eastern Siberian tribes. They are widely spread from the west Yenisey River to the Sea of Okhotsk, from North Sakhalin Island and the base of the Taimy Peninsula in the north, south to Lake Baikal, the Amur River and northern Manchuria in Mongolia. Two related tribes to the Evenk are the Even or Lamuts found on the north east coast and the Nanay in the lower Amur Riverbasin. The Evenk have a huge Evenk Autonomous Area, Okrug, north of Krasnoyarsk which covers 745,000kms^2, but this is only a fraction of their total homeland. Their total population currently numbers 30,000, though only a maximum of 1000 are still living through nomadic reindeer herding. Their language is related to Chinese although they are culturally closer to the Mongolians. The name Evenk means 'he who runs swifter than a reindeer' and there is a strong sense of superiority within this aristocratic tribe.

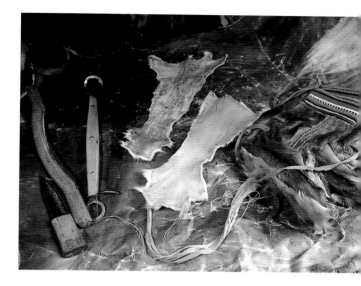

▲ The tools used for making skin footwear, (left to right): knife, skin-softening tool, two-ended scraper, tanned reindeer shin skin, tanned shin skin after softening, back tendon sinew threads, and the finished boot.

◢ Evenk winter boots consist of two parts: an inner boot with the fur inwards, and an outer boot with the fur outwards. Between the two, an insole of dry grass or sedge is used. It took Anna five days to sew these boots together, using no pattern but that which she carries in her head.

Surviving 65°C Below

When the Siberian winter arrives you need to be well prepared. Clothing for the extreme cold has evolved over many generations. Not surprisingly, the answer is well-tailored clothes made from wool and fur.

Evenk Prehistory and History

The Evenk originally came from Central Asia, drifting north over 1000 years ago away from the hostile Huns. Evenk mythology includes many descriptions of plants and animals which have never been found in Siberia and many costumes of shaman priests showed representations of animals never found in the north. As a group they can be traced way back into Chinese history, a great deal earlier than any of the modern European nations. The earliest records of the Evenk are 11th century BC.

Somewhere in the prehistoric past in south eastern Siberia and north Manchuria, the Evenk first domesticated reindeer and began to use them as mounts. They then gradually spread out across Siberia. Reindeer riding increased their efficiency both in hunting and during hostilities. A good sized reindeer can travel 80kms a day with an 80kg pack over terrrain that would be far too rough for a horse.

Historical information on the Evenk is rare. The Russians first discovered them in the 16th century, living what they termed a 'primitive' life. Contact with the Russians intensified in the 17th century first in the west and then east. As the Russians penetrated their area, the Evenk retreated. In these early days the Evenk pleaded with the Russians to limit the encroachment of their hunting grounds to protect their economy and culture. The majority of ancient Evenk songs of that time told of unhappy love, poor hunting and disease. Three quarters of the Evenk population died during the 19th century.

In those early days of Russian migrants, the Evenk would make traditional clothing for the large influx of miners into Siberia who suffered from the intense cold. They also sold them birch bark baskets, berries and mushrooms. Furs were traded with the Russians in exchange for tea, ammunition, flour, sugar, salt, tobacco and alcohol.

The Living Conditions

There have been people living in the taiga forests for the last 6000 years, yet historically it has had a very low density of population as these huge primeval forests are an exceedingly difficult habitat in which to survive. There is a basic lack of fat and carbohydrate available, which is essential in a place that experiences such low temperatures. Food is scarce and animals only have fat on them in the autumn. In comparison to the Inuit, the Evenk do not have a good supply of fatty food to keep them through the winter, nor is

If you are used to solid winter boots made in western factories, Evenk footwear seems hopelessly lightweight. But their soft tanned fur boots allow plenty of room for warming foot movement. They are breathable, allowing perspiration to pass away from the foot; and without a solid sole, they reduce the conductivity of warmth to the cold ground. Made in several parts they can be easily aired at the end of the day. They are extremely comfortable to wear.

Full fur clothing is the only way to stay warm when deep in the taiga tending to the reindeer herds. Gloves are permanently attached to the jacket sleeves with a palm slit that will allow fingers freedom to tie knots and other intricate work.

The woman's clothing here consists of several layers of heavy wool. Although the clothes are extremely functional, they are beautifully embellished.

Making a Salt trough

there a ready supply of vegetables and fruit as an alternative. However, unlike the Russian migrants there are no known cases of the Evenk suffering from scurvy or rickets and this nutritional balance can only be maintained by eating between two and four kilogrammes of meat each day, some of which is raw.

The Evenk Nomadic Lifestyle

The traditional Evenk live in brigades made up of four or five families, with as many as 1400 reindeer between them. Truly nomadic, they pack up all their possessions once a month in order to find enough grazing for their reindeer herds. According to them, 'the best thing in life is leaving for the next place'. The concept of ownership of land was unknown to the Evenk. A brigade territory would usually centre on a river and include the land either side of it, the boundaries were usually flexible and usually land would be shared by more than one clan. There are usually six or so regular sites used, depending on the availability of timber and water as well as grazing. Along their migratory routes, the Evenk would make permanent storage facilities so that seasonal clothing and equipment could be stored when not in use. It was accepted that other clans might borrow items from these caches and return them after use.

The Evenk lifestyle is one of hunting and herding, though activities pertaining to reindeer predominate and in terms of raw materials, the reindeer provide them with most of their needs, meat, milk, leather, clothing, shoes, shelter and transport.

Hunting, Trapping & Fishing

The taiga forest is home to a large number of indigenous species including moose, wild reindeer, elk, roe deer, bears, wolves, wolverines and many species of fish. Wild fowl include wood grouse, ptarmigan, various geese and ducks, many of which are hunted and trapped for wild meat. Some species are still hunted for fur as the majority of the Evenk income is raised through selling pelts. Sable, polar fox and squirrel are shot or trapped and silver fox, mink and muskrat may be farmed. A husband and wife team could spend several months alone in winter in order to trap sable. The sable meat is given to the dogs.

Infants sleep securely tied into heavily-insulated wooden papooses. During the day they are left suspended in these ingenious cradles from a low branch of a tree, lulled to sleep by the gentle swaying of the breeze. They also travel in these papooses, strapped on to the side of a pack reindeer. In case of accidents their cradle is equipped with a roll bar.

Reindeer Herding

Maintaining the reindeer herds is a full-time occupation and at times during the year when extra labour was required, larger groups of Evenk and their herds would come together.

To protect their herds from possible attack by predators, such as wolves and bears, dogs are used. If the herds need coralling the Evenk are skilled at making large enclosures by rough cutting wood in an extremely short period of time. Another threat to the herd is that posed by wild reindeer who will either take away domesticated does or mate with them so they produce part wild and part domestic stock. These offspring were comparatively slower though stronger than fully domesticated reindeer and some Evenk groups would encourage this. Others actively discouraged it saying that mixed stock reindeer could never be tamed, that they had a tendency to attack people and would endeavour to return to the wild herd. The reindeer fawns which arrive in April and May would share the does' milk with the Evenk.

During the brief summer period, June to September, the herds feed in the forests and may be moved onto the treeless tundra to be fattened for the winter. In winter, from early October to May, reindeer paw through the snow for lichen or 'reindeer

▼ Here, Sasha makes a salt trough by felling a tree, propping it at a convenient height for the reindeer and then hollowing it out with his axe.

▼ The reindeer are herded into the forest clearing where the trough has been constructed. They naturally congregate around it for the salt, offering the herders plenty of time to study the herd and capture any of the normally flighty reindeer that they need to.

Birch Baskets

moss'. The lichen grows only a few millimetres a year, taking over 30 years to reach maturity. The animals trample and compact the snow over the lichen as they feed hence the need to keep the herds on the move and continually find new grazing.

The Evenk recognise individual reindeer by their body markings and faces and those that are to be trained or killed for meat are caught by lasso. Reindeer that are selected for riding or pulling sledges ·are carefully chosen and then castrated before training begins. Not only can they carry large weights, over long distances, reindeer are also particularly suitable for pack animals in the taiga forest as they are small, neat and able to make there way through the dense cover. They are ideal for winter conditions as they have broad hoofs that do not fall through thick snow. Riding saddles would have been made from carved bone or wooden framework with cushions stuffed with elk hair. A rug of elk and bearskin is thrown over the saddle and another is used to protect the pack from the elements.

Wild Plants

Very few wild plants are used although a number are edible. Birch bark was used extensively in the past and is still used to

Arkadi rolls the birch bark he has gathered to scrape off the accumulated crusty lichens. The outside of the bark will be the inside of his basket, cleaning it will make a better basket and enable him to fold the bark more easily.

Inside the chum, Sasha sews the now cut and folded basket with thread. Above them reindeer shin skins are drying. On the stove a loaf of oma bread is baking.

The finished basket will be used for the transportation of food. Arkadi has made hundreds of these in his lifetime.

Bark containers and berry gathering baskets. Bark baskets folded and sewn in this way are far easier and quicker to make than many other forms of basketry.

make containers. Berries are collected, often by the children, using a method that efficiently fills the container as it is skimmed over the top of the bushes.

Shelter

The Evenk live in conical tents known as 'chums'. Closely resembling a tepee, as a shelter this concept has been used worldwide for over 70,000 years. The chum takes only 15 minutes to put up. The dwelling area is first cleared of snow and then covered in spruce twigs for insulation. The frame is made with approximately 30 young larch poles that are stripped of branches and would have been covered with 160 or so reindeer skins in winter and birch bark in summer. Now canvas and tarpaulins are used. A single chum provides shelter for two or three families.

Inside the central area of the chum is a fire place or cooking

Building a Chum

Inside a chum the smell and the atmosphere has changed little for thousands of years. But the topic of conversation has, now the politics of the outside world are reaching them. The twins are may be one of the last generations to experience the Evenks' nomadic lifestyle.

area. Larch is the preferred fire wood a. birch gives off too much smoke, thoug stoves and modern cooking pots have replaced the open fire. The side areas o the chum are used for sleeping, and are first laid with birch or larch branches and then covered with reindeer skins. A reindeer skin sleeping bag, made up from eight hides, is all that is required to provide the warmth and insulation to survive winte night temperatures down to -60°C.

Tanning and Clothing

Modern style clothing is favoured during the summer months, but the traditiona dress made from reindeer hide is still used to survive the winter months. Both men and women wore coats, leggings and thigh high boots. Unlike other Siberian indigenous groups, the Evenk wear their coats short to allow them to be worn when riding. A cotton outer layer is sometimes worn over the breathable reindeer clothing which protects the outer layer and stops the formation of ice due to condensation.

With no suitable vegetable fibres available in the taiga, reindeer sinew is used to make thread, by drying, twisting and then joining it into long pieces. Like the caribou sinew used in the Arctic, this thread will outlast the garment itself. Their method of tanning hide is quick and highly efficient. Rotten larch wood is used as a tanning agent as it containes sulphanates which soften, dye and preserve the hide. The powder is first mixed with water and then rubbed into the scraped skin, the skin is then folded and allowed to 'lie' for 24 hours. Tannin found in bark may also be used as a tanning agent. The softening of the hide is carried out equally efficiently using a piece of wood and a simple twisting action. All items of clothing, including footwear, are made up by the women. Boots are made primarily from the skin from reindeer legs as the short hair is particularly durable. Reindeer socks are worn inside boots with the fur innermost giving a double insulating layer.

Reindeer hide would also be used to make nappies for babies. Placed in a basket, the hide is lined with rotten larch wood, which acts as a highly absorbent liner.

Building a
Chum

Sledges and Skis

Every self-respecting Evenk man must be capable of making a sledge of his own. Several different types are made depending on whether they are for men, women, or for carrying loads. In the dense taiga forest sledges are used less frequently due to the nature of the terrain and also because reindeer as pack animals are preferred. Larch wood is used for both sledges and skis, it grows extremely slowly but it is the most durable timber in the taiga. Taking a week to complete, the sledge's dimensions are slight but it is immensely strong, lasting several seasons. Tools used include the bow and strap drill and whittling is carried out using a knife sharpened on one side only with the user moving the wood, not the knife. Tools are always made with wooden rather then bone handles as bone becomes impossible to hold at extreme temperatures. In addition to the axe, another vital tool to the Evenk is a knife which is always kept on a belt, these might have been decorated with bones and amulets to protect the owner against evil spirits.

Harnesses for sledges and ski bindings are made from reindeer hide. With a complex system of knots and ties the straps and ropes are designed to take the strain and withstand the extreme winter climate. Decorative work used to adorn reindeer and sledge harnesses has been made of mammoth bone that has been discovered buried in the permafrost. Some groups of Tungus used mammoth bone to make a calender, with notches

1 The basic framework of a chum is made of three poles lashed together. Against this another 10 poles are laid.

2 Covers used to be of animal hides, but today, old canvas cloths are lighter and more convenient. Laid over the poles and tied on at their corners so that they overlap, they give protection from insects, wind, rain and snow.

3 To exclude drafts the base of the chum is packed with moss. Most chums have a thin sheet metal stove and chimney for heating. It may be snowy outside but by Evenk standards the interior will be comfortably warm in comparison. However, unless you have been born to this life it still seems cold.

Collecting Berries

corresponding to the days of the year which was divided into 13 months.

The skis themselves are made by splitting a piece of larch wood into planks and then shaping them with an axe. The ends of the skis are heated over a fire so they can be bent upwards. Two different types of skis are made. On the first, strips of reindeer calf skin are glued to the underside of the ski with the fur facing backwards. The glue is made from boiling fish skin, usually perch. The layer of fur prevents the skier from slipping backwards on the ice. With the second type, the wood used is selected for its high tar content which acts as a priming wax. The design of these skis is almost identical to a pair found preserved in Scandinavia which have been dated at over 5000 years old.

Food & Preparation

Reindeer is the most important food source and all edible parts of the animal are used to ensure that everyone receives a full nutritional balance. It is because reindeer meat is so lean that such a large daily amount is required. To obtain the vitamins and minerals that reindeer meat cannot provide, the Evenk eat the part digested contents of the reindeer stomach. The entrails, liver and kidneys are all eaten, boiled or raw and the blood is drunk. Finely cut pieces of raw

If there is a vital principle of survival it is 'to achieve maximum efficiency for the minimum effort'. No better example exists of this than the Evenk berry collecting basket. Swept with a deft flick of the wrist through billberry bushes, the berries are brushed into the basket without damaging the bushes. In 20 minutes a bucket can be filled with these delicious berries.

Too big for the papoose, but too small to ride alone this child is tied onto the reindeer saddle for safety while travelling.

Like generations of Evenk before him Arkadi relaxes in the taiga whilst gathering birch bark. Despite the turmoil of Russian politics the fundamentals of life in the taiga have changed very little.

The future for Siberia

A helicopter brings supplies to the small Evenk community of Surinda. Flights are much less frequent than in Soviet times, when the state airline did not have to worry about paying for its costs.

A few Russians also inhabit the taiga. Like the Evenk, they are also finding life more difficult. Yuri, a Russian sable trapper and hunting guide, lives in the taiga from the end of September to late February but the market for sable fur is in serious decline.

frozen fish also provide essential nutrition and other foods are prepared from imported flour. A typical meal might include reindeer cheese and strong Russian tea.

The Future

With over 200 generations of Evenk people living this traditional nomadic lifestyle, it is likely that this current generation will be the last to do so. Maintaining this way of life has become increasingly difficult, balancing it with modern interventions and with pressures exerted by oil and gas extraction and associated pollution problems, the loss of land, a recognised need to educate their children and increasingly less assistance from the State.

After the Russian Revolution in the 1920s, the government provided the Evenk people with a monthly income and reindeer. The reindeer were then owned and traded by the government. A helicopter service was provided that would provide regular supplies and could fly people out to purchase provisions and sell pelts. This was free of charge. They also provided pensions, medical help and schooling. The children would attend boarding school for nine months of the year, only returning home for the summer months.

Between the 1930s and 1950s, the Soviets also pursued a policy of 'Sedentarisation of Nomads'. Many native villages were established and the women, children and elders were encouraged, or coerced, to settle. The men continued to herd and hunt. Now many villages have been consolidated into larger settlements with the male population still herding away from home or joining the increasing number of unskilled and unemployed workers. Reindeer meat is still the mainstay in their diet and the women collectively tan hides and make clothing for outside markets.

Since the fall of the USSR, those still living a nomadic lifestyle now receive little as the payments and perks have been stopped. They have to pay for the helicopters, few medical supplies are provided and many children remain at home.

The extraction of Siberia's vast mineral wealth has had its price to pay and many say that the only thing that has protected Siberia from total ruin is its vastness. Hideous pollution problems such as polluted rivers and leaky oil pipelines have severely effected the indigenous people as they complain not only about loss of land and polluted grazing, but that their staples of fish and reindeer meat taste of oil. Their livelihood depend on the bounty of the tundra which cannot now be guaranteed. Many still wish to continue living one of the most harsh and hard-working lifestyles, asking for only a handful of modern luxuries. As this last generation leaves the forests and Siberian wastes, with them will go the wealth of accumulated knowledge gained from living deep within the taiga.

Oxana has not yet been to school but has already started to read and write. Evenk children are educated at state boarding schools, but leaving their community is a wrench.

The taiga is one of the emptiest places I have ever visited: the atmosphere is that of a house the day after a great party. If you ever are lost in the taiga I hope you have the fortune to meet the people who choose to live there; you could not find more able and gracious hosts than the Evenk.

Western

The Western Samoans know the price of paradise. Relying on ancient skills, they walk a shoreline tightrope: making their living from an ocean that during the cyclone season threatens to engulf them.

Samoa

Like jewels in the ocean, the islands of Western Samoa are an earthly paradise. Their beauty is matched by the people who inhabit them and whose joyous singing is perhaps an echo of the ancient voyagers who made landfall here. The Western Samoans make their living from their coconut plantations, their incredibly fertile gardens built on volcanic rock and from the reefs and the ocean. They still work closely together as once their ancestors must have done searching for habitable land.

The Western Samoans call the coconut the tree of life. This long elongated nut is the sennit coconut, the fibres from the husk are used to plait exceptionally durable cordage, an essential part of the mariners equipment.

Western Samoa is a far cry from the Pacific island paradise of white sandy beaches, swaying palm trees and coconut cream as portrayed on film or in the travel brochures. To the traditional island inhabitants, the Western Samoans, whose way of life, or *fa'a Samoa* has been adapted to harvest all the available fruits that exist on tropical islands with little surface water, harsh currents, razor sharp corals and few land mammals, it is no paradise at all. Quite simply, you cannot survive on the view.

The idyll of the South Pacific has long been an inspiration to writers like Jonathan Swift and Daniel Defoe and the many film makers who captivated their audiences with romantic notions of being a castaway in the bountiful south seas with exotic flowers and friendly inhabitants. Robert Louis Stevenson actually spent the last four years of his life in Western Samoa, which he chose, not for its beauty, but for its regular postal service, courtesy of New Zealand.

Geographically, Western Samoa belongs to the Polynesian Islands which form a triangle between Hawaii, Easter Island and New Zealand and include The Cook Islands, Tonga, and Tahaiti. In distance, Western Samoa is 2700 miles east of Sydney and over 2600 miles southwest of Hawaii. It consists of two main land masses, Upolu and Savaii and two smaller islands Apolima and Manono lie in the strait between the two larger islands. The islands are volcanic with a narrow coastal plain and rocky mountains that cover the interior; Mount Silisili, the highest mountain reaches 1850m. With a total area of less than $3000kms^2$, the coastline is 403kms in length, most of which is coral reef. A typically tropical climate, Western Samoa lies squarely within the South Pacific's notorious typhoon belt and it experiences devastating typhoons, on average every ten to fifteen years.

Culturally, Western Samoa belongs to the Samoas. Originally one nation, American and Western Samoa are now a homogeneous nation, politically divided. They speak the same language, practise the same customs and to a greater degree pass on the same traditions. Compared to American Samoa (a territory of the United States), independent Western Samoa is considered the poor relation. It may be materially less well off, but culturally it is a great deal richer. The larger of the two,

Non-chiefly men, denoted by their banana leaf necklaces, prepare giant elephantear roots for cooking. This has replaced taro as their staple food, since a mysterious disease destroyed their crops after the cyclones.

Building a Hut

An interior of the roof of a fale shows how the frame work of hibiscus poles is used to support the thatch. Traditionally cordage was used to tie these frames together, today nails serve the same function.

Western Samoa is comparatively a quieter and more gentle country which nurtures the most traditional Polynesian society in the Pacific Islands. Although it has moved into the 20th century, it has done so without haste and many of the rituals and tribal hierarchies remain, almost unchanged, despite 200 years of colonisation.

Prehistory

There are a number of theories that have been put forward to explain the advance of early man throughout the widely spread Pacific Islands. One assumption is that the Polynesian people entered the Pacific from the west – the East Indies, the Malay Peninsula or the Philippines. This theory comes from the fact that lapita pottery, similar to that found in New Caledonia, was found in the Samoa's and Tonga (dating from 1500BC to AD) but no use of pottery was found when the Europeans first made contact with Polynesia in the 17th century. Another theory, proposed by Thor Heyerdahl, suggests that the Polynesians migrated from the Americas and this is based on the presence of the sweet potato in the Pacific and South America, but not in Asia. Samoan legends support the theory that the Samoa's were initially settled by the Fijians or Tongans, with the earliest known evidence of human occupation in the islands being tentatively given as 1000BC.

Building a fale is a group effort – everyone turns out to lend a hand. In this way few jobs take more than a few days to complete.

How to build a
Hut

European Contact

By the early 16th century Balboa had discovered the Pacific Ocean, and in 1521 Magellan navigated his way across it. Although some whalers, pirates and escaped convicts landed in the Samoan islands at earlier dates, the first European to record them officially was the Dutchman, Jacob Roggeveen. He sighted the Manu'a Islands (now American Samoa) in 1722 while searching for the Terra Australis Incognita, the great unknown southern continent. He gave the islands Dutch names and than sailed on without landing. In May 1768, the French explorer Captain Louis-Antoine de Bougainville passed through the islands and seeing the islanders travelling about in ocean-going canoes, christened the archipelago 'Les Iles des Navigateurs'.

Later in the century a number of Samoans were killed following raids on European ships. This gave them a reputation for being hostile people and the traders therefore, steered clear of Samoa until the early 1800s. By the 1820s several Europeans had settled in Samoa, most of them escaped convicts or retired whalers who were welcomed by the unsuspecting islanders. Some of the first itinerants to land there introduced a form of Christianity and later in the 1800s, the missionaries arrived. The Christian gospel was readily accepted by the Samoans and has been an integral part of life to the present day. Between 1850 and 1880 many European settlers arrived primarily for the purpose of trade. They established a society and minimal codes of law with the consent of the 'Upolu chiefs, who still maintained sovereignty in their own villages.

Until the end of the last century Samoa was ruled by chiefly families, often in temporary unions. Western Samoa was finally annexed in 1899 as a colony by Germany and was then taken by New Zealand in 1914. Between 1920–1946 it was a mandated territory of New Zealand under the League of Nations and from World War II, a U.N. Trusteeship administered by New Zealand in preparation for self-government. In 1962 it was granted independence and became a member of the Commonwealth.

1   The roofing is made up of split coconut fronds.

2   The split fronds are cross woven to make effective waterproof shingles.

3   The woven split fronds are tied onto the roof so that they overlap each other.

On the Reef

Society and Culture

More than any other Polynesian people, the Western Samoans are guided by tradition and closely follow the social hierachies, customs and courtesies established long before the Europeans arrived. Beneath the surface lies a complex code of traditional ethics and behaviour. A sense of mystery surrounds the outwardly friendly Samoan people. The traditional *fa'a Samoa* is closely guarded and well preserved, especially in the remote villages where European influences are still minimal.

A *matai*, or chiefly system, is in effect throughout the islands. Each village comprises a number of *aiga*, or extended families, which include as many relatives as can rightly be claimed. The larger the *aiga*, the more powerful it is and to be part of a powerful *aiga* is the goal of every traditionally-minded Samoan. The *aiga* forms the base of Samoan society and all possessions and money are shared within it. Family belonging is essential and if anyone breaks away and lives alone it is regarded as a tragedy. Family business is open and includes all aspects of love, sex and marriage. Society is male dominated, with elders seen as natural leaders and extended great respect.

An *aiga* is headed by a chief, or *matai*, who represents that 'family' on the *fono*, or village council. The *matais* can be male or female and they are elected by all adult members of the *aiga*. The influence of a chief is measured in the provision he can make for his people. The giving and acceptance of gifts by all is most significant within Samoan culture. The *matais* are also responsible for law enforcement and deal out punishment for any infractions that occur within a village. Social protocol is taken extremely seriously and there are harsh punishments for crimes such as adultery and violence.

The highest chief, or *ali'i*, sits at the head of the *fono*. Other titled people include the talking chiefs, or *tulafale*. The *tulafale* is an orator who will carry out ceremonial duties and liaise between the *ali'i* and outside entities such as foreign visitors. The symbol of the *tulafale's* office

▼ **The Western Samoan knowledge of their reefs and treacherous currents is exceptional. Strong swimmers, they gather much of their food from the sea, using, in this case, an eel trap.**

is the *fue*, which looks like a fly wisk and is made from a four kilogramme mop of coconut fibre, or sennit. It represents wisdom and must be carried whenever the *tulafale* is in an official capacity. A *tulafale* also carries a staff which, with the *fue*, is handed down through the generations. The remainder of the village, the untitled members, are encouraged to interact only with their particular group, for example unmarried women or schoolchildren.

An obvious feature of Samoan culture is its peaceful nature. Conflict is usually resolved through negotiation as open conflict is considered rude. To many Europeans, Samoan society is considered romantic but somewhat primitive, the Samoans in turn regard ours as powerful but also primitive.

Fire

Many of the traditional skills that have allowed the Samoans to thrive so successfully in their tropical island habitat are still in use today and remain an integral part of their culture. Indeed, the making of fire with a fire-plough is still taught to young schoolchildren as an essential survival tool. The fire-plough is used to rekindle the fire, which myth says, is stored in trees having been conveyed from regions deep beneath the earth.

Shelter

With lush forest covering most of the islands, including broadleaf and evergreens, vines, tree-ferns and many epiphytic plants, there is no shortage of materials to make the variety of shelters and living quarters required. Although stone has been used for building, in particular for Christian churches, many of the more remote villages have returned to using traditional construction methods in the aftermath of the devastating typhoons 'Ofa and Val which struck the islands in 1990 and 1991.

The type of building materials are chosen depending on the purpose for the shelter. A camping shelter, *fale laufai*, would be make of banana leaves and a temporary shelter or even cook house, *fale laupolo*, from soft wood and light thatching. The most substantial would be the living quarters, a traditional open-sided

The eel trap comprises a box with bait fish inside. There is only one way into the box: through a tube fashioned from an old tin can, and attached to the end of this is a sleeve of cloth. The eel enters the box in search of the bait, pushing through the cloth, but is then unable to find a way out. The box is placed on the reef with its entrance facing the coral, not outwards from the reef. It is then covered with old pieces of dead coral to both weigh it down and camouflage it.

Fruits of the Forest

hut with a roof and supporting posts, which are fashioned from hard wood and built to last. Although today nails are used, in the past sennit fibres, made from the fibres of the coconut husk, would have been used to tie the timbers together. The Samoan family required shelter from sun or rain for cooking and living and the dwelling houses needed to have an even damp-free floor as people slept on floor mats. This is provided by a layer of dead coral. Free circulation is obtained by open-sides and protection from the wind and rain by the moveable coconut leaf wall screens. These also act as a protection from typhoons as open-sided dwellings are less prone to wind damage. Extra wind protection to protect the thatch is given by placing coconut fronds on the ridge poles.

The ceremonial dwelling, *fale talimalo*, built in a similar style to the dwelling house, provides shelter not only for meetings of the *fono*, but also for the *kava* ceremony. This is the traditional focus of all life and theatre of Samoan power and provides a public opportunity to display rank and dignity. The kava plant, *Piper methysticum*, is cultivated and used to make a drink which is the centre of this ceremony. The root is crushed and dried to a

The leaf of the barringtonia plant, from the seed and bark of which a fish poison can be extracted. This was used to catch fish in both inland streams and coral heads. Today, with conservation issues better appreciated, this practise has been abandoned. Other environmentally damaging techniques are also falling into disuse as their effects are appreciated.

The leaf and fruit of the Great morinda. This plant is very commonly found on the beaches of Pacific islands.

powder which is then mixed by a young woman from the village, typically a virgin. This is then served by untitled young men in a coconut shell cup. The *kava* cup can only be accepted in a particular order related to the rank and standing of the ceremonial participants.

Fruits of the Forest

Cultivated plants are grown in plantations inland of the coastal village. Samoan horticulture has never been intensive with people growing enough for their own needs and a little extra for festivals. The key tools of the past would have been the digging stick, planting stick and bush knife. Fruits like oranges, mangos, avocados, passion fruit and soursop are seasonally obtained, with papaya, bananas, guavas and pineapple cropped all year. Pele leaf, cabbages, tomatoes and many herbal and spicy plants, as well as yam, taro, plantain and sweet potato which are the most important tuberous plants for giving high carbohydrate yield, are all consumed. Sugar cane is used for sweetening and its leaves for thatch and pandanus for matting. Cordyline or ti was originally used to make everyday clothing and the inner bark of the papermulberry tree for the ceremonial bark cloth, *siapo*. The women in particular still have extensive knowledge of the local fauna and its uses in their traditional medicines.

▼ **Forest fruits are gathered in baskets woven when needed from coconut leaves.**

◄ **The great morinda fruit, or 'stinking cheese fruit' has the texture and smell of strong brie. Before it reaches this smelly stage it is a reliable emergency food. Some Samoan communities depended on this fruit after devastating cyclones.**

Dug-out Canoes

Single outrigger dugout canoes are a way of life for Western Samoans, and they use them for fishing both inside and outside of the reef. At night they fish well outside of the reef using a kerosene lamp made from an old milk tin fixed to a wooden stick, the wick fitted into the tin lid is made from a wad of dry leaves.

The 'Tree of Life'

Of all the plants found within the islands, the coconut provides the Samoans with the most in terms of raw materials and it is known as 'the tree of life'. Climbing a coconut palm is achieved with ease using a climbing bandage that is wrapped around the ankles, originally made from a loop of bark. All parts of this plant are utilised, the young and mature leaves, the flowers, fruits, stem and roots. Food, implements, shelter, containers and clothing such as water bottles, nets, mats shark rattles, fishtraps, thatching, sandals, tongs and drinking cups to name but a few.

Hunting

These remote islands have never been home to many land mammals apart from the flying fox, small sheath-tailed rat and the Polynesian rat and then to feral pigs. Pigeon hunting was carried out in the past when the bird population could support such activities; the bow and arrow was abandoned for warfare and used only for fowling. This lack of land protein forced the Samoans to utilise their most valuable asset, the reef, and to harvest the bounty available they developed a number of complex and ingenious techniques to trap, stun and catch the variety of shellfish, fish and marine animals.

Canoes

Waterborne transport was originally by canoe, though history tells of stories of incredible long distance swimming feats

The confidence and courage of Samoan fishermen knows no bounds. This fisherman is using a rattle made from a forked stick and dry coconut shells. Shaken in the water the sound of the rattle attracts sharks which he then snares.

How to make a
Canoe

between islands. Seven types of canoes were made with the size and number of outrigger booms being determined by their uses for speed, distance, paddling or sailing. Today, only the dug-out canoe, or *paopao*, and the bonito plank-canoe survive, both with outriggers and propelled by paddling. The timber comes from a variety of trees and for a dug-out, once a suitable trunk has been identified it is cut down, roughly shaped and then hauled back to the village for hollowing and the making and placing of outriggers, booms, floats and connecting pegs. Bailers would have been made of coconut shell and lashings from sennit braid. Men took canoes beyond the reef for deep sea fish such as bonito and flying fish as well as dolphins and sharks. Fishing nets made from coconut fibre, in particular the sennit coconut, are still used and the men continue to practice their legendary breath-hold diving to great depths to spear fish.

The Reef

The waters within the bounding reefs provided the main source of fish and these waters would be combed over and over again in every conceivable fashion from groping between the rocks with bare hands to using skilled devices such as traps, nets and lures. The Samoan's great knowledge of the habits and movements of the various species of fish influenced invention and method and these ranged from individual to community efforts. Sennit rope nooses, coconut root lures, scoop nets, futu fruit poison, funnel-shaped basket traps, turtle shell hooks and even cobwebs. Women were never allowed to take canoes,

1 A suitably sized tree is felled. As much of the shaping as possible is carried out in the forest, using a chain saw and adzes. With the rough hull shape and most of the hollowing finished the boat is carried back to the village beach for completion.

2 With the fine hollowing of the canoe finished, outrigger supports are then lashed on using plaited sennit rope. This has more give than nails or screws and is therefore stronger at sea.

3 The outrigger is simply fashioned from a piece of wood carved to cut easily through the water.

Octopus Lure

instead using baskets in the lagoons to catch fish which they attract either by artificially-created rock heaps, traps, spearing or by drawing octopus, for instance, out of their coral holes with sticks. Both men and women need to be adept at floating over the sharp coral with their faces under the water to watch for fish but avoid being bowled over by the waves and strong currents.

Men taking canoes beyond the reef would have used natural methods of navigation such as the stars, wind and the morning dew. Flocks of sea birds which pursue schools of small fish would alert them to large groups of bonito fish, and a whole fleet of canoes would follow them with trailing hooks. Ocasionally the canoes would get so full of fish that the fishermen would jump overboard and swim their canoes to shore.

One of the most enjoyable communal fishing activities still practicised today, is the *lauloa*. Each family will collect and shape a set length of coconut leaf and vine netting. These leaf sweeps are then joined together to form the *lauloa* and taken out to the breakers with people of all ages lined along the entire length. The *lauloa* is drawn on a falling tide and with great laughter and shouting the fish are driven towards the net. The catch is then divided between the participants.

Western Samoa, though now wholly part of the modern world, has managed to retain a number of these customary practices and they are used daily by many people. A tropical south sea idyll, it cleverly conceals the vast knowledge and intricate skills that have been developed by its inhabitants that ensured they survived and thrived in their island habitat. The beautifully carved warclubs and the papermulberry cloth may now only be produced for a visitors market, but the traditional values and *fa'a Samoa* are still fundamental to Western Samoan society and culture.

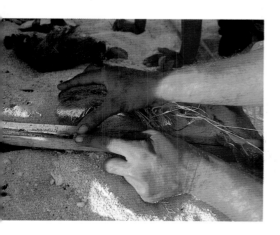

The traditional method of firelighting here is the fire-plough: one piece of hibiscus wood is rubbed against another in line with the grain. This skill flourishes in Western Samoa.

A simple lure fashioned from a rock, some leaves and sennit rope. Made to look like a rat it is used to entice octopuses from their holes in the reef. Once an octopus grasps it, the lure is reeled in and the octopus caught by hand.

Namib

The Bushmen's way of life is one of survival. Relying on their ability to track animals through the dusts of this waterless landscape, they wrest their living from one of the Earth's most hostile wildernesses.

Despite its richness of game and birdlife, Bushmanland is a harsh environment in which to survive. Dust baked dry by the merciless sun, it tests all living things to the limits of their survival, including people. The Bushmen who inhabit this land maintain skills and knowledge that can be tracked back to the earliest dawn of our species. Treading lightly in the thorny scrub they strive to maintain a balance between the complicated physical and spiritual forces that shape their world.

The bushland of north eastern Namibia is one of the most difficult places on earth to live. The bushmen and women that make their home here have developed exceptional skills and a close rapport with their environment.

The hunter gathers of the Kalahari sands in Southern Africa are known as the Bushmen or San. These gentle people are recognised for their superb tracking ability and great daring and resource as hunters. Legendary too is their ability to find water in areas that might go for a year with little or no rain.

Events over the past three millennia have meant the San were pushed out of a large percentage of their original range, the more accessible areas of southern Africa, now known as South Africa, Botswana, Angola and Namibia. The Bushman stronghold became the more remote and desert-like expanses of the Kalahari. The current population of the San people is on the increase, so there is little danger of the race becoming extinct, but their traditional lifestyle is likely to become so, maybe within a decade. Today, probably none lives totally as a true hunter-gather and most are living as subsistence farmers. However, the traditional skills have not all been lost and those who live in areas where they are allowed to hunt, will supplement their diet with food sources from the bush.

Of the Bushmen or Khoisan people living in southern Africa, divided into the San and Khoi people, there are three separate groups. The northern group consists mainly of San or !Kung people who live in the area to the west of the Okavango Delta in Botswana extending east to the town of Kasane, north into Angola and west into Namibia. The Ju/Wasi or Ju/'hoansi people belong to the !Kung group and they are found in Eastern Bushmanland, around the town of Tsumkwe or Tjum!kwi, in north east Namibia.

Origin of the Bushmen

Evidence suggests that the late Stone Age, which began approximately thirty thousand years ago, only ended in the Kalahari region about 200 years ago. Anthropologists agree that the early Stone Age people are the ancestors of the modern Bushmen making them the product of more than a million years of human evolution in Africa. It is likely that prior to 3000 years ago the Khoisan were the only occupants of southern, central and parts of east Africa and were widely

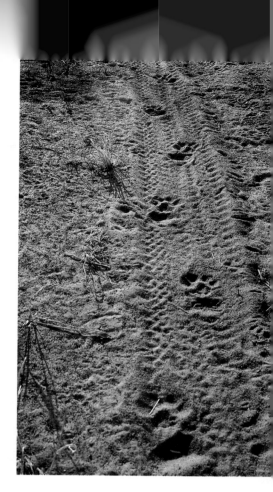

The dusty ground in the bush has enabled the bushmen to develop the skill of tracking to the highest possible pitch. Here a lion has walked down a bushroad after a car passed by in the early evening. Being able to recognise the tracks of large predators is of course a fundamental skill. Being able to accurately determine when the track was laid can be a more important factor, one which takes much practise to acquire.

Each bushman hunter carries a hunting bag that contains his essential tools for survival. He travels incredibly lightly, enabling him to track animals for many miles.

Poison Arrows

⚫ **Gao and N/nani making arrows under the shade of the village leadwood tree.**

dispersed across the continent. Highly adaptable, they survived by hunting and gathering and had a highly developed social structure and artistic skills. Bantu groups moved in from several different directions around 3000 years ago, bringing with them the knowledge of cattle-rearing, pottery-making and iron-smelting. The San would have traded with them for iron, copper and salt in exchange for skins, ivory and horn.

Since that time the San have competed with pastoralists for territory and they have been ousted by warfare, annihilation and absorption. Confirmation of absorption comes from evidence found in several of the Bantu languages. San languages are the only ones that originated with the distinctive 'click' sounds, for example '!' which is made when the tongue is pulled away from the mouth, to create a vacuum. Reproductions of this and other clicks are found in the Zulu, Xhosa and Tswana languages which indicate the process of intermarriage and absorption over time.

The ⚫ Bushman arrow is made with a detachable foreshaft, so that the injured animal is not able to pull out the poison-bearing head with its teeth.

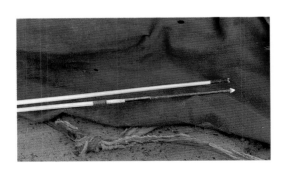

How to make
Poison Arrows

Contact with Europeans

Further evidence of clashes with pastoralists are found in the records of the early Cape settlers. The white settlers who arrived in 1652 did not take kindly to raids on their cattle by the San and they were declared as vermin and therefore hunted to their death. It is suggested that as many as 2500 Bushmen may have been killed during the last decade of the rule of the Dutch East India Company at the end of the 18th century and many others were taken prisoner. The San were not easily beaten and in addition to taking livestock they would plunder homesteads and killed a number of settlers. In some cases bloody battles were fought.

As the Bantu and settler populations increased, the San found themselves inhibited in their movements by settlements on the margins of the Kalahari. Dependence on these people grew with the San becoming herders themselves as well as working for the cattle owners as their previously extensive nomadic territory was being eroded and game was getting more scarce.

History from 1950s

Bushmanland covers an area of over four million acres of wilderness, semi-arid bushland with scrubby tree cover, towering baobab trees, seasonally wet pans and open plains. The western area of Bushmanland is largely uninhabited, with the only main settlement Tsumkwe in the east being the administrative centre. It is around Tsumkwe that the current population of 2000 Ju/Wasi live, a population that has seen a number of major changes in the past few decades.

Interest in this area and its people was first generated in the 1950s by a film maker John Marshall and his mother Laura, an anthropologist. At that stage some of the Bushmen were still relying solely on their hunting and gathering techniques, though

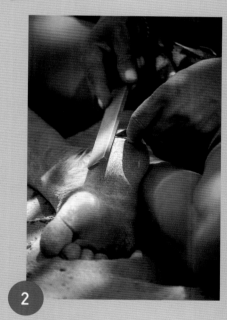

1 The Bushman arrowhead is fashioned by cold hammering a piece of fencing wire.

2 The arrowhead is filed into shape and sharpened to a razor edge.

3 Poison, the juice from the chrosomelid beetle larvae, is daubed on the shaft of the arrow. The point is deliberately left poison-free to prevent the hunter accidentally poisoning himself.

Among the Elephants

most were subsistence farmers supplementing this with the traditional methods. Namibia was under the South African government and by the 1960s South Africa was pursuing its homeland policy and had set aside areas for ethnic people, dictating where these rural groups could live. As a way of repressing their nomadic ways, many of the areas were not where people would have chosen to be or indeed could even sustain them. This was combined with the setting up of local facilities in Tsumkwe, a shop, medical facilities, a school, farming advice and welfare payments. A pattern seen with so many indigenous groups, the population of Tsumkwe swelled as the attraction of western clothes and money attracted the people in from the surrounding villages. More than 1000 Bushmen were eventually living in the town. All the surrounding area was gathered and hunted to nothing, the people had little to occupy themselves, spiritually they were deflated and alcohol and violence became a major problem.

At the beginning of the 1980s Marshall returned to discover the plight of the people from the old villages. He and others worked to establish what is now called the Nyae Nyae Foundation to encourage the people to return to the villages and back to the land. The foundation and the Nyae Nyae Farmers Co-operative are organisations that work for and with the San to promote and develop projects for health, agriculture and education, environmental planning and wildlife management. The sustainable utilisation of resources including tourism are now being promoted to generate awareness and income.

Sniffing the air to identify the intruder, the Namibian elephant has the reputation for being not only the largest but also the most aggressive of all the African elephants. Weighing up to six tons, even a landrover provides no sanctuary from their wrath if angered.

Travelling in groups, elephants graze the sparse bushland leaving a trail of devastation, behind them. They may seem cumbersome but they actually move with remarkable stealth and agility.

The average African elephant requires 100 litres of water each day. This makes them a fierce competitor for water. Boreholes supplied to the bushmen by the government have to be encircled with boulder walls to prevent the elephants accessing the water and pushing over the wind-operated pumps.

Making Fire

Gao blows an ember to flame in a bundle of dry grass. Fire is the fundamental focus of Bushman social life. The twinkle of a Bushman fire glimmering in the black bundu night symbolises the continuation of a way of life that has not changed for thousands of years.

The Bushmen Way of Life

Originally purely nomadic, the San would move with the seasons to optimise on any available food sources. In an environment where water is extremely scarce, times of drought might bring more than a dozen months without rain and temperatures in the height of summer easily exceed 95°C. Likewise, on a clear cold winter night the temperature will plummet to below freezing and frost is not unusual. The San share their environment with a number of predators who would not hesitate to prey on the San themselves.

The number of people supported through hunting and gathering is directly related to the available resources. When game and plants were plentiful, the San would congregate in large groups and as resources diminished, the groups would fragment into smaller units, even down to a single man, his wife and immediate family if required. It was a way of life where self-reliance was emphasised more than interdependence between the larger group members. Bushmen are opportunists and if hunting was unsuccessful they were known to take kills from other predators, even from lions.

The San originally had no organised system with either chiefs or a headman. This type of system would have been seriously flawed given the regular amalgamating and separating of different groups. All decisions were therefore made by group consensus. This lack of leadership or political system led early settlers to pronounce that the Bushmen were of course a primitive people. In fact their system was ideally suited to their harsh environment where the promotion of good human relations was essential for the survival of all. There is a strong need to feel part of a group without hostility, so people will freely speak their minds which avoids the build-up of tension. Personal possessions were few due to the constraints of a nomadic lifestyle, but these would be gladly lent with permission and would bind the lender and borrower together.

Hunting

In Eastern Bushmanland only the Ju/Wasi are allowed to hunt by the traditional methods. Firearms are not allowed, which at times when game is extremely scarce means they have to rely on bows and arrows, hunting on foot and without the use of vehicles or dogs. Hunting game was and still is the domain of men.

A traditional hunting set would comprise a skin bag containing a spear, bow, a digging stick and a quiver. The quiver would contain arrows, fire-making sticks and sometimes a drinking straw. The bow and spear are made from branches of the raisin bush and the bow from a branch of grewia bush. Each would be cut, debarked, cut to the right length and then treated with animal brain or marrow to give some water proofing. The bindings would be made of sinew and the bow string from long strips of twisted sinew from the back muscle of an antelope.

Spear tips were made from a piece of iron and then burnt into the shaft with the joint strengthened with sinew. Spears are used to kill larger game species, weakened by poisoning. Two types of arrow, the reed arrow and branch arrow are used, the former has arrowheads made of iron, fencing wire, horn or bone for game and the latter uses has a bone arrowhead for birds. Both will break on impact with the animal so that the arrowhead separates from the main shaft to ensure that the poison on its shaft remains in contact with the animal. Both parts of the arrow can be retrieved and reused. Bushmen bow and arrows are small and the killing of game relies on the use of poison. There are a number of natural poisons available, most are in the form of body juices that are squeezed out of insect

Gao carries two means of starting fires: (left) flint and steel in a small tin made from an old battery and (right) a hand drill friction set.

Using a hand drill friction fire set, made from either marula, baobab or mangeti wood, Gao and N/nani start a fire.

Setting Snares

pupae found buried deep in the sand. The poison is placed on the shaft of the arrow head only, so that the hunter will not be cut and poisoned himself by the tip. An antidote is carried, but no antidote is available for a full dose. The poison is deadly, killing its victim in any time between one hour and three days.

A hunt would have been organised between a small band of men, between two and five individuals, who would search for game to feed the whole group. The actual hunt will begin once the quarry had been sighted, initially their tracking skills will identify and follow the fresh spoor of a potential prey species. Through expert knowledge and experience Bushmen are able to interpret much information from examining an animal's spoor, its size, age, weight, sex, whether that animal is in calf and therefore its likely behaviour.

Once an animal has been tracked and stalked an arrow will be fired and if it successfully hits the animal, the San will follow it, though only for a short distance to establish the line of flight and that the arrow has become embedded. Following it immediately will only cause it to run further. They will then return and track the animal later to find it dead or dying. Meat around the arrowhead is discarded and the more perishable parts such as the liver may be eaten immediately. The remainder is placed in carrying skins taken back to camp and distributed according to strict customs. Bushmen seldom kill for no more than their immediate needs and very little is wasted. The meat actually belongs to the owner of the arrow and not necessarily the hunter that fired it. The meat first goes to the hunters and is then divided between their closest relatives. An excess of meat would seldom, though occasionally, be gorged. It is usually saved and dried for another day. Entrails would be cleaned and preserved and even half-digested grasses squeezed for their water.

The Ju/Wasi now hunt mainly small buck like steenbok or duiker or larger species such as kudu and frequently guinea fowl and bush pig. In the past, larger species such as giraffe, now illegal, would have been taken. Snares and traps are often used for birds and smaller game species, the design and mechanism depends on the intended prey species. They are set up where fresh spoor indicates a frequently used path. Rope for the snares was made from the long silky threads of the wild Kalahari sisal that were

This guineafowl snare has been made from sansaveira fibres and set in the shade of a small bush.

The trigger of the trap is set finely baited with a gum or with a seed that the guineafowl is tempted to eat. Reaching for the bait, its head will pass through the noose which is triggered by the bait being moved.

plaited and spun into various lengths and thicknesses.

Digging sticks or poles would have been used to hunt around in holes in the sand for spring-hares, badgers, ground squirrels and porcupine. Porcupine were particularly sought after, with the hunter actually entering the burrow. Having blocked up all the remaining entrances, he would crawl into the burrow and literally blockade the creature in its hole. From behind a protective shield the hunter would grab its legs, then the neck and throttle it.

The Bushmen are known for their elaborate camouflage efforts and the ability to use very little for cover when stalking an animal. A favourite was mimicking an ostrich using a feather frame on the shoulders and the real head of a dead bird attached to a pole. The hunter would then behave like an ostrich, pecking at the bushes and meanwhile have his bow and arrow ready as he stealthily approached a group of birds. The San did not think that ostriches were particularly intelligent and one story says that when raiding an ostrich nest they would always leave one egg behind just in case the ostrich forgot what it was doing and did not lay any more eggs.

Fire, Water and shelter

Whereas the Bushmen villages now have taps and borehole water, in the past as much as 90% of their water would have been found from non-standing water sources. Water when found would

▼ **For the Bushmen, the drought means that not only is game more scarce and difficult to hunt but also that it is difficult to travel the long distances required for fear of dehydration. It had been three and a half months since even the most expert hunter had made a kill.**

▼ **Wildebeest rush away from a waterhole at the arrival of the Bushmen. With few open waterholes available, nature is harsh, forcing prey and predators into close proximity. Survival requires sharp reflexes and excellent sensory awareness.**

Telling your Future

Divining is a popular pastime. Using bones or charmed sticks a diviner predicts the future and diagnoses the cause of current events. He will stack the sticks in a set order, rub them over his head and armpits, blow onto them then suddenly slap them down onto the mat in front of him, asking a question of them. By their relative position he can determine the answer.

have been stored in ostrich egg containers filled, sealed and then buried. They would also be used as water bottles when travelling. Some hollow trees stored water well into the dry season and tsamma melons, which belong to the cucumber family and ripen during the dry mid-winter, were sought after. The flesh of these fruits is 90% water and it is said a man could survive on one alone for six weeks. The greatest source of water was the succulent roots and tubers that would be gathered then pared, grated and squeezed to provide water.

In some areas the conditions favour the construction of 'sip-wells', often made and used by the older women and by hunters. Particularly damp areas are identified by the presence of certain plant species and a hole will be dug as deep as possible into that area of sand. A hollow reed or grass-stem is inserted into the hole and the lower end packed with grass or roots. The sand is then replaced and tightly packed and trampled. After an hour or so, the woman returns and draws heavily on the reed straw. This reduces the atmospheric pressure and draws water away from the sand grains and up the hollow tube. Water may then be transferred to and stored in an ostrich egg shell.

As nomads, the shelters would have been simple, constructed from grass and branches, and still preferred by some of the elders. As subsistence farmers, the circular, wooden framed hut with mud sides and a thatch roof is more common and more typically now, corrugated iron huts in the larger settlements.

Gathering

A greater proportion of the diet is produced by the women during gathering trips; on average 60–80% of the total food consumed. They may set off two or three times a week on foraging visits and return with 15kg of vegetable material as well as nursing a child. Traditionally the women would have taken

The sticks are carved to specific shapes and patterned by scorching. They comprise two pairs representing male and female forces with the fifth stick representing an undefinable force.

The Bushmen love singing, it is common both night and day to hear the gentle sound of Bushman voices. Accompanied by the thumb piano and a rythmic clapping, the sound is beautifully haunting.

A bees nest in the hollow in this baobab trunk has been raided many times for its honey. To make the access easier the Bushmen have constructed a ladder of pegs hammered into the trunk.

carrying bags made of springbok, duiker or steenbok skin and decorated with ostrich eggshell beads and used wooden digging sticks. A dried tendril in an otherwise parched landscape is all the women would need to indicate the presence of the water-bearing tubers, hidden deep beneath the Kalahari sand.

A wide variety of vegetables, fruits, berries, honey, roots and tubers were collected during the changing seasons. The key tool in preparing food would have been a mortar and pestle often carved from ironwood. It was used to pound the tsamma melon seeds to make a flour, the various nuts and berries and to pulverise dried meat for small children.

Tanning and Leatherwork

Skins of hunted animals would all have been used to provide items of clothing and pieces of equipment. The tanning process

Gathering Party

🔺 **Women scour the bush for edible plants: hard work that requires sharp eyes to spot the withered stems of edible tubers.**

involved soaking the dried hide and rubbing it with a mixture of animal brains and fat and burying it with dung and then leaving it for two to three days. The piece of skin would then be hammered between two pieces of wood and finally softened with the hands. Various colours of hide were achieved by using different plant dyes. Sinew provided the thread and it was sown with an awl.

The loin cloth originally worn by the men is still used by some people today for hunting, as it allows the hunter to move silently and much faster than modern clothing does. Made from a piece of duiker or steenbok skin it would have been decorated with glass or ostrich eggshell beads. Likewise were the decorated dancing skirts made from springbok skin and used by the women and girls. These would be worn over their loin apron. The skirt was the traditional dress amongst the Bushmen women and they are still worn today for some ceremonies. Loin aprons are still worn by some women under western clothing.

Carrying bags and hunting bags were made from skins. A whole skin, preferably gemsbok, wildebeest or hartebeest would be used to make a carrying skin or cloak. This was used for many purposes including carrying gathered materials, household goods or children. Tied around the body and across the forehead, heavy loads could be carried for long distances and the skin would also double up as a blanket in cold conditions.

The Bushmen are known for their craftsmanship of ostrich eggshell beadwork which was used to make jewellery as well as decorate items of clothing. The eggshell would be broken into small pieces with the fingers, stones and sometimes teeth. Holes

🔻 **The contents of the hunters bag. From left to right: an axe, a digging stick, a quiver with poisoned arrows, fire sticks, a knife, a flint and steel and a file for sharpening.**

This tuber is both edible and thirst quenching. Split open and the core mashed to a pulp, its juice will be squeezed out into this infants mouth to quench his thirst.

Digging for Grubs

N/nani is searching for poison grubs. Digging 80cm from the base of this *Comiphoa africana* bush he will excavate a hole 40cm deep, sifting out the grubs from the sand as he digs.

Inside this hard shell is the orange grub he is seeking. He needs many of these grubs as each arrow requires the application of poison from ten grubs to be made sufficiently deadly.

are drilled in the centre of the bead by turning the 'drill' in between the palms of the hand. The beads are then strung onto sinew and each bead individually chipped until it is round using a springbok horn and a stone. The string of beads would then be polished using a grinding stone. Early trade links with Bantu people in the past introduced glass beads which were often used in the place of ostrich eggshell.

Stories, Songs and Musical Instruments

The myths and tales of the San are many and they vary greatly from area to area and from group to group. Many of the stories tell of the origins of aspects of their environment. They may have moral overtones and form the basis of how they see the world. Generally they reflect the good humour and humanity of these people. The existence of the stars is explained in such a way. Long before the existence of the San, a young girl of an ancient race wished for some light in the darkness of night so that she might see her way home. Leaning down, she took some ashes and threw them high into the air, there they stayed and became the Milky Way. The Bushmen say that when stillness falls at night and even the geckos are quiet, you can here the stars calling.

Music and singing are all important to the San. The San will make and play music on the music bow and thumb piano for pleasure around a fire and use the piano whilst walking long distances. With keys made of fence wire and attached to a board of wood, the thumb piano is held in the palms of both hands and played with the thumbs.

The Trance Dance and Traditional Medicine

Dancing is used on many ceremonial occasions to give thanks for a successful hunt, and most notably the trance dance. This is a spontaneous event involving a number of dancers where one or more may reach a trance state and effect healing by the laying on of hands. The dancers will gather around a fire and one of them will start to sing. The volume increases as more voices join in and the women clap rhythmically from their fireside seats. One man will rise to dance and the others join him, stamping the sand as the pace quickens and the tension and energy increases. The dancers may wear dancing rattles around their legs which are made from moth cocoons strung onto a piece of leather and filled with small stones or pieces of ostrich eggshell, which adds to the rhythmic sound. The fire is fed and the combination of the men's song, the clapping and the women's wailing and shadows dancing allows the dancers to slip into their own world and maybe one or two men will slip into a full trance-like state. During this time the spirit world becomes visible to them and the dancers will appeal to them with passion and even violence. These visions are described in detail in many of the Bushmen paintings found across southern Africa. If the aim of the dance is to heal a member of the group the healer will draw this affliction into himself and then 'hurl' it out with a shriek as the 'evil' leaves his body.

The San believe that a powerful healing force lies in most people and that it is effectively activated by the trance dance. This trance condition can be dangerous and those who acquire it will sometimes collapse into the fire and have to be rescued from the burning embers. To the San the trance state is very important as it allows them to meet the spirits and their survival in the living world is dependent upon the successful communication with the spirit world.

Other methods of healing are found in traditional herbal medicines concocted from local plants, an immense knowledge of which has been developed over the millennia.

A baobab nut broken open to expose the edible citrus-like pulp inside that tastes like freeze-dried ice cream. This food is nutritious and can be stored for long periods if the nut is undamaged.

Inside the baobab pulp are many seeds. These are roasted in hot sand before being ground up and used to infuse a coffee-like drink.

Surviving the Drought

The Future

It is sadly a function of change that sees the two essential requirements for the traditional Bushmen way of life disappearing at an increasing pace. The number of cattle living in and around the Kalahari has put huge pressure on the natural resources: what they graze the game animals cannot. Cattle continue to move into the remote areas and the number of game animals has decreased. Without space and without the plants and animal species that they depend on, the San cannot survive in their traditional manner. Beguiled by the material attributes of western society, the San have no choice but to follow them as the alternative is no longer viable.

Hopefully the rudimentary aspects of their traditional culture will be retained by those people to whom hunting and gathering is not only a way of supplementing their existence, but also a way to retain the links with the past. No doubt some of their skills and craftsmanship will survive and provide a sought after income from visitors to the area. It will be a sad time though if the spiritual aspects of Bushmen life and the energies of the trance dance are eventually found sealed in Bushman art alone.

Despite the constant encroachment of outsiders on their land, the Bushman way of life continues. Perhaps the greater threat is from the drought that threatens the very possibility of their traditional way of life.

Drought has been a constant problem for the Ju/hanse. This is a sip well excavated over centuries by Bushman hands digging down in search of water.

Like the spirit of a Bushman cave painting leaping into reality, the silhouette of the hunting party returning to the camp speaks of the continuation of a way of life that reaches to the earliest dawn of our species. It would be a tragedy to lose the living knowledge these people possess.

Indone

The jungle can be a forbidding and hostile environment to the uninformed. The people who inhabit these regions are themselves daunting to meet, but it is only from them that you will discover how to live there.

sia

The island of Seram is covered in dense tropical rainforest and plantations. Sheer razor-backed hills reach down to the coast clad in giant buttressed trees, rattan, lianas and bamboo. The Nuaulu, with a fierce reputation for headhunting, are experts at obtaining food from the jungle and they make living in the rainforest look easy. Their most important tool is their parang, wielded with great expertise they use this to fashion traps, shelters, bows and arrows and fishing baskets.

Cacanene the village shaman shows us a damar resin lantern. Made from leaves filled with powdered resin; the Nuaulu explain that they learned to make them from their ancestors. These lanterns are the only form of lighting permitted inside their clan ritual houses.

Lying between the Seram and Banda Seas is the old Spice Island of Seram. Mountainous and heavily forested, the rainforests of the interior were the original home to the Nuaulu people. Traditionally animist in their beliefs, the Nuaulu still thrive on the materials and foods found in the rainforest.

Indonesia

An archipelago with more that 13,000 islands, the Republic of Indonesia stretches for nearly 5000kms from the Asian mainland to the eastern part of Irian Jaya in the Pacific and 1700kms north and south over the equator. Most of the islands are mountainous, some dominated by active and extinct volanoes. The climate is typically tropical with heat and high rainfall giving humid conditions all year. The natural vegetation for nearly two thirds of its area is tropical rainforest.

With its incredible variety of natural resources, Indonesia attracted people from all over the globe to trade its crops of coffee, tea, gold, pepper, sugar, tobacco and teak. Trade with Asia had already been established by the first century AD and mention is made of the islands in Ptolemy's writings of 165AD. By the 16th century battles were being fought over spices from the Moluccas or Spice Islands, now known as Maluku. Cloves, nutmeg and mace were so rare and expensive that they lured navigators from far and wide. The Dutch arrived in Indonesia in 1596 and ruled for 350 years, though some islands remained independent. Full Indonesian independence was finally granted in 1945.

The rugged terrain and island culture gave rise to a wide range of people with separate languages, cultures and religions; ethnic groups number over 300 with more than 350 languages and dialects. Most are of Malay stock and descended from people who originated in China and Indochina. Most of the remainder are the darker skinned Melanesians found in eastern Indonesia.

Seram

The Maluku's second largest island is Seram. With its mountains and forests, there is an air of mystery to Seram. It is known as Nusa Ina or Mother Island as the Molukans believe that this is where the ancestors of central Maluku came from. It is part of the biogeographic region of Wallacea, a term used to designate those islands that lie in the transition zone between south east Asia and Melanesia, Australia and beyond. It was named after Alfred Wallace who wrote *The Malay Archipelago*, one of the most important natural history books of the 19th century. The larger islands of the Malukus, are still dominated by tropical rainforest, although most of the smaller ones are now denuded and extensively planted with clove, nutmeg and coconut palms, especially along the level coastal land. The total population of Seram is around 250,000. It is populated mainly by Malays and transmigrants from Java and Sulawesi. Seram is home to a number of indigeous people.

Prehistory and History on Seram

It is likely that there was some form of human presence on Seram as early as 40–60,000 years ago. The earliest examples of Holocene flakes are from around 8000 years ago and there is evidence that metal and other trade items, such as Chinese ceramics, had come from mainland south east Asia around 2000 years ago. The Neolithic and pre-Neolithic period lasted up until 300 years ago and would have seen the arrival of the wild pig, goats and the domesticated dog.

Seram had significant contact with Java, Sumatra and mainland Asia before the 15th century. Early Muslim influence from the sultanates of Ternate and Tidore, other Spice Islands, was apparent from the 12th century onwards particularly at the

The jungle is not for the arachniphobe. In an environment where insects abound their predators thrive too.

A teenager is adorned with a red cloth headdress as part of his tribal initiation.

Living in the Jungle

Fresh drinking water can be found in rattans and liannas. A section a metre or more long from a large vine can give over a litre of water in good conditions.

eastern and western ends of Seram where there were trading networks. The civet was introduced at this time for its musk was used as a perfume fixative and probably the cassowary from New Guinea. Portuguese missionaries arrived in the 16th century and in the early 17th century the Dutch established a number of trading posts, ruling the island from about 1650. Deer were introduced, possibly by the Muslims, as late as the 17th century; contemporary wildstock seem to be descended from tame animals imported from Java and Sulawesi.

There were at least 19 indigenous Seramese languages and these remained fairly separate until the 17th century when the disruption caused by the Dutch attempts to control the spice trade and later colonial attempts to consolidate power, resulted in large population movements and a mixing of languages. Many local languages were substituted by the Ambonese dialect of Malay, a process that is still continuing.

The human population of Seram show many similarities in the mode of subsistence. The role of the sago palm is central for the production of edible starch, construction materials and utensils. Pottery is historically of limited importance, most containers are made of sago spathe, bamboo or basketry and more recently coconut and calabash. The cultural map is dominated by the Alune to the west and Wemale to the east. In central Seram, the dominant highland interior indigenous group were the Nuaulu.

The Nuaulu

The Nuaulu are now a people of south central Seram and during the past 25 years have occupied five settlements along the narrow coastal strip in the old Muslim domain of Sepa. In 1971 the population was 500 and formed half the Nuaulu speaking population of the island. The remainder are located on the north coast at Seleman Bay. There are now approximately 2000 Nuaulu people located around Sepa in six villages.

Until the 19th century, Nuaulu clans occupied separate small clan hamlets in the highlands; hamlets with a separate ritual house and dispersed menstruation huts on their periphery. At that time there had been relations with the Muslim coastal settlements for around 200 years and there were intermittent periods of animosity including head-hunting and warfare, alliance and trade. In west and central

Chewing betel nut and smoking are popular Nuaulu pastimes.

Seram, head-hunting was still practised until several generations ago.

Towards the end of the 19th century, the Nuaulu began to occupy sites near to Sepa, due to pressure from the Dutch and the coastal rajas and their coastal resettlement policy. This movement persisted into the 20th century and was significant in that it changed the Nuaulu's environmental and economic conditions and hence their knowledge and use of local fauna. However, they are still significantly oriented to the mountainous interior in their use of natural resources rather than to the coast. It is only recently that there has been a reversal to this coastal movement, when the government opened up inland areas for resettlement. This has allowed some Nuaulu to return to highland areas from where they were previously evicted.

Nuaulu Society

Before moving to the coast, a single autonomous clan, or *ipan*, would occupy a single hamlet. The *ipan* still retains considerable independence in ritual, political and economic matters, but now many of the villages contain five or nine separate clans. Each *ipan* is divided into two *numa*, descent groups, and is focused on a single ritual house. The clan is headed by either an *ia onate*

A long strip of bark is being pounded to soften it for use as a ritual loincloth.

Hunters use both spears and bows with long unflighted arrows for hunting pigs and birds. Their accuracy with these weapons is not to be underestimated.

Processing the Sago

ipan or *kapitane*. Clans are theoretically equal in their status and the clan headman is known as the *Matoke*, or in Ambonese Malay as the *tuan tanah* or lord of the land.

Their traditional religion, animism, focuses on the worship of ancestors. The veneration and circulation of sacred objects and spirit mediumship are an important part of this. These practices have survived conversion to Islam and to a lesser extent Christianity. Mortuary customs vary, but usually the corpse is left exposed, wrapped in matting and lodged between tree branches or on a bamboo bier.

Both men and women take part in initiation ceremonies. For the women this comes at the time of their first menstruation when they move to live in separate menstruation huts on the edge of the village. The longer they spend there the greater the accolade received by the girl's father. The adolescent boys take part in a complex initiation ceremony. Before the ceremony, permission is asked of the ancestors to go ahead with the procedure. The boys still wear the bark cloth loin cloths, they stand on prepared log platforms and a cuscus is killed for each participant. The most important aspect of the ceremony is awarding each boy with his

▼ **A mature sago tree is felled. The soft tissue of this tree is packed full of edible starch and forms the staple of the Nuaulu diet.**

red headdress, which he will wear for the rest of his life. All adult men wear them and the cloth is so symbolic that it may be used as currency. It is also important how the men wear their hair under the turban and this style will reflect different phases in their lives.

Nuaulu Way of Life

The forest is the Nuaulu's way of life. It is the main source of animal protein, much of their vegetable food and it supplies the materials needed for construction and other technical purposes. The forest provides the environment to support hunting, trapping, fishing, swiddening and gathering, whether for themselves or for cash crops. The Nuaulu have little knowledge of the coastal environment which is close to the village, beyond the immediate rocky and sandy shoreline. Some gathering and fishing is carried out, but their habits have little changed from their interior upland roots.

The most important forest product for the Nuaulu is metroxylon sago which produces their starch staple. Swiddens are cut each year. These are energy intensive to maintain and

The troughs and guttering necessary for processing the starch from the sago fibre is improvised from the sago tree itself.

The trunk of the sago is split into two neat halves using a wooden wedge and a rock maul.

The starchy fibres of the sago are cut out from the trunk using a bamboo adze. These are then washed and a pure white starch settled out in a complex arrangement of troughs.

Cooking in Bamboo

Split and splayed, out a fishing basket is woven from a single piece of bamboo. An old basket is used as a mould to weave the new basket around.

In use the basket is held with its opening upstream while the leaf litter and river vegetation is agitated to disturb freshwater shrimps, crayfish and crabs which are then swept into the basket.

produce food bulk, but contribute a small amount to the total diet in terms of food value. The crops cultivated are varied but are dominated by manioc, taro, sweet potatoes, yams, bananas and plantains. Prior to World War II, most cash was obtained by the Nuaulu through the collection and trade of forest products such as dammar resin and honey, but since Indonesian independence in 1945, the Nuaulu have become increasingly involved in the growing of clove and coconut palms for copra. Level coastal land is heavily cultivated with cloves and coconuts. Even the steep valley walls of the short rivers descending to the sea are cultivated. Animal domestication for food is virtually non-existent though some chickens are kept and dogs are used for hunting.

Rouhua

The Nuaulu village of Rouhua has a population of approximately 300 with five separate clans; the number five is symbolic to Rouhua as is the number nine to other villages. The area from which they extract forest products is approximately 900kms^2. The village is mainly animist. All children attend the school in the neighbouring Christian village, though most drop out and the literacy rate is low. The loss of young people from the village to date has been almost negligible, as there is little else for them to do other than subsistence farm.

The Forest

As with many traditional people, the Nuaulu conceptualise the forest as a total experience and they are highly skilled and knowledgeable about how it can provide for them. Their existence depends on these living skills. The greater part of the Nuaulu forest environment consists of mature forest and this is differentiated into two contrasting types, *wesie*, which is forest unaffected by human activities and where there are no land rights, and *wasi* where land is cultivated and land rights do exist.

The Nuaulu always venture into the forest with diffidence, taking both practical and magical precautions. Travel is largely by established paths, using well-known landmarks for direction-finding such as rocks and prominent trees as well as artificial markers. Habitual paths are subject to rules of upkeep which apply to routes between villages and gardens. At some strategic points there are forest huts, *nma wanane*, resting places for long distance treks.

The Nuaulu recognise over 700 individual plant species of which over 200 have specified uses. These may be for sacred purposes, used in rituals, for primary and secondary food sources, medicines, beverages, building materials, clothing, manufacture of tools, ornamentation, resin, firewood, and colouring for scents and cosmetics. All parts of the plants are used: the resin, fruits, leaves, flowers, bark and branches as well as whole trees though ritual restrictions may apply to certain products at certain times.

Swiddening – Crops and Gardens

A swidden is generally a maximum of four kilometres from the village and less than a hectare in size and before a new area is cleared, permission must be sought from the ancestors. Therefore a five-log platform is erected, offering tobacco, a shaving from a gold ring and some betel nut.

Where possible, forest is cut close to a watercourse and adjacent to cleared plots to allow easy access. This usually takes place between December and January, allowing time for clearing, firing and planting before the heavy rains begin in April and May. Work will begin at the centre of the plot so that timber can be felled uphill, to allow downhill clearance unhampered by

1   To cook the fishing haul a section of large bamboo is filled with the catch, in layers separated by several manioc leaf dividers.

2   A little water is added from a bamboo carrying container. Only enough is added to create steam, not enough for boiling. The end of the tube is then plugged with a large wad of manioc leaves.

3   The bamboo is set on a fire made from dead bamboo started with a bamboo fire saw. Rotated occaisionally it cooks over a fierce heat for about 40 minutes. Despite charring on the outside the bamboo does not burn through.

4   Once the cooking is complete, the bamboo cooking tube is deftly split open onto a banana leaf revealing the orange of the beautifully cooked shrimps and cray fish. A meal caught in bamboo, cooked in bamboo over a bamboo fire lit with bamboo burning bamboo fuel.

How to cook a Wait

Building a Hut

⊕ **The most fundamental tool for jungle life is a parang. The Nuaulu value sharpness greatly. Outside of every permanent or semi-permanent shelter you can find a sharpening stone set upright by wooden pegs.**

fallen stands with the largest trees taking up to half a day to fell.

Once cleared, the swiddens are then planted. In the first six months tubers such as yams, sweet potatoes and taro are cropped and in the second year onwards fruits and vegetables such as papaya, bananas, soursop, durian and plantain are planted and in older gardens, manioc. Cash crops of coffee and cloves are also planted. Gardens are also kept in the village, with maybe a few fruit trees grown close to the houses.

Uses of Timber

The characteristics of individual woods are well understood by the Nuaulu. Their properties, hardness, strength, susceptibility to disease and rot, elasticity, resilience, ability to split, and quality as fuel are recognised and therefore a species is carefully chosen for the task at hand.

Forest cutting activities are an entirely male domain, though the number of people involved may vary. Generally it is carried out in singles or pairs, though larger parties, clan lead, may be assembled for purposes such as cutting timber for building ritual houses. The main tool for all forest activities is the long bladed parang. Iron blades have probably been in use for more than 500 years and before they would have used Neolithic stone blades.

The transport of timber is almost entirely manual, mainly carried on the shoulder by individuals or pairs of men. Women

1. To make a shelter, the ever useful fast growing bamboo provides the building material.

2. In only an hour the basic framework of a shelter can be built. The frame is lashed together with the outer fibres from a bamboo pole. Very strong, these shelters may survive over a year.

3. The roof is made from thatched palm leaves. The leaves are sewn individually onto battens using a thread from the fibres of a palm frond. Under this, a raised floor is constructed to keep the men out of the way of biting insects when relaxing.

4. The flooring is made from split bamboo. A large pole is split at its nodes every 1.5cm all the way round and then split so that it can be unrolled like a blind. Then the inner part of the nodes is trimmed off (here) so that it will lie flat.

may occasionally carry small amounts for firewood in back baskets or on their heads. Timber may be stored in a temporary shelter or *kamrenta*. Wood for items like shields is generally rough-cut and stored in houses.

As a village may contain more than thirty dwellings and ritual houses, the Nuaulu's main use for timber is for house frames. Timber is also required for timber stores, drying platforms, stages for male initiation ceremonies, menstruation huts, copra drying huts and ground level houses. In addition they build temporary shelters such as garden huts, hunting huts and huts for the coconut season.

Other uses for timber include domestic utensils, spear shafts, stakes, bow and arrows and hooks. Soft woods are used for short-life cooking utensils, and hardwood for heavy-duty objects such as barkcloth beaters and mortars. The most significant items are the large drums, or *tihane*, of the *suane* ritual house. Types of firewood depend on whether fastburning or slowburning materials are needed. Domestic hearths are ideally kept burning continuously. This is collected by women and the older men on the immediate periphery of the village.

Other non-edible forest products include dammar resin which is collected as a cash crop as well as for burning in the ritual houses. No other source of light is allowed in the ritual houses. Rattan is used in basketry and as a cash crop on

After a small ceremony to bless the hut, it is ready for use, with ample room to shelter at least six people from a torrential downpour.

How to build a
Hut

Forest Fauna

demand. It is also used to make ties for constructing ritual houses, as nails are not allowed in these structures. Bamboo is used in numerous ways including basketry, firewood, utensils and firewood. Dyes and poisons are extracted from the various herbs, climbers and roots. Several different species of tree, usually fig species including the banyan, are used to make the bark cloth. Most of these products, with the exception of resin and rattan, are found in close proximity to the village.

Sago Processing

Of primary importance is the sago palm, *Metroxylon sago* and each household will use around 12 mature sago palms every year. The sago palm occurs naturally in the forest, though they are also planted by the Nuaulu. The sago palm reaches maturity within 15 years at which stage the pith contains the maximum amount of flour. Sago extraction and processing is an exclusively male activity and sago possesses a high cultural value, equal to that of meat. The sago palm, once processed, is the main source of their starch staple and it is being processed continually.

The palms are hacked down and a broad strip of the bark is taken off the trunk to expose the white pith. This pith is then adzed to break it up before it can be processed. The processing equipment is made each time a palm is cut down using the palm,

⊗ These little beauties are palm grubs. They munch their way through the trunk of fallen sago trees. A Nuaulu delicacy, they are eaten whole or by splitting them open and sucking out the inside. Actually they taste quite pleasantly of palm oil. They can also be roasted over a low fire.

⊗ Damar resin is gathered for sale and for use in ritual. Lanterns made from this are the only internal lighting permitted during ceremonies.

bamboo and coconut matting. The process is carried out near a stream as running water is required. Water is poured over the pith which is kneaded to dissolve the starch. The starch is then collected as it is deposited and the excess water trickles away.

When boiled with water, sago forms a thick glutinous mass or porridge which is eaten with salt, limes and chillies. Sago bread is made from the dried, powdered, sifted and baked raw sago. It is hard and dry and will keep for months. The conversion rate of this food resource is extremely high, with two men finishing one sago tree in a few days they will provide food for one household for a month.

Once the sago has been harvested, some pith is left behind in the trunk to allow the cultivation of the sago weevil. The larvae are eaten either raw or cooked, and are a good source of protein. The remains of the plant are used to provide building materials and tools. The sago leafstalks are used for the walling in houses and the sago fronds for the thatch.

Hunting and Trapping

Hunting is carried out exclusively by men. They only use bows, arrows and spears. The Nuaulu do not use poisoned arrows. They did have access to guns for a short period, but these are now banned. Dogs are used, either singly or in groups, with each

Pain is written across this mans face. He is very slowly pulling the comb from a wild bees nest in this bankside. The bees have not been smoked and his only protection from their stings is a plastic bag over his hand and his red headdress pulled down over his eyes.

Jungle Foods

household having an average of three. The dogs are fully domesticated and will be selected as hunting dogs when puppies. Dogs are well cared for and cannot be killed except on the payment of a fine. Even a wounded dog will be allowed to die a natural and 'good' death.

Hunting may be carried out by individuals or in groups. This may be a small group of related people or a large group of about 30 people who are providing for a specific event, often for a ritual obligation; this is normally on a clan basis. For ritual events only the meat from deer, pig, cassowary and cuscus is eaten. Hunting trips may extend over several days. On extended hunts wives and children may accompany the hunters. Such expeditions may reach considerable distances from the settlement, over 25kms, though normally they are confined to a radius of 6kms. This form of hunting is often undertaken at night.

During such hunting trips and when in the forest cropping coconut and sago, the Nuaulu rely purely on forest products for subsistence. The shelters made are constructed with bamboo and a prayer to the ancestors is placed in the thatch on

▲ **The Spice Islands live up to their name. This is a nutmeg growing wild in the rain forest. The fleshy case is very tart but popular eaten with lime juice and sugar.**

▼ **Almost no source of protein goes unnoticed. Nuaulu hunters climb deep into the bowls of the local caves in search of bats which they consider excellent eating.**

completion. Water is sourced from the sap of vines and creepers and is drunk as long as the liquid is clear and not coloured. Originally fire would have been made using a fire saw, though this technique has been lost to lighters and matches.

There are five different types of trap the Nuaulu use, one of which is the leg hold trap which may be used for pigs or even bush turkeys. The spring-loaded spear traps for pigs and deer and made from bamboo and are lethal. They are quite capable of killing a human. After setting one, the setter will always leave a 'sign in the bush' to indicate the site of the trap to prevent an accident. The cleared forest and gardens actively attract pigs and deer and as they are not fenced, traps are often sucessfully laid around the margins.

Wild pig is the most important source of protein to Nuaulu, providing 30% of the total meat consumed, it is hunted by spears, bows and arrows and traps. Pig butchering is very ritualised and done in a particular way as with the distribution of meat, for example, an offering is first made to the ancestors. Little attempt has been made to domesticate pigs though, except they will fatten piglets if the sow is killed. Occasionally, trapped pigs will be kept or traded. Pig grease is a focal symbol of the Nuaulu identity, particularly as many Moluccan Muslims fear ritual contamination from anything connected with pigs. The men will smear their bodies with pig grease for certain major rituals.

Deer are also hunted and trapped, providing 12% of meat eaten. The hides are used to make drumskins and the young antlers and craniums are used to make househooks. The large antlers are sometimes sold as they are an ingredient in certain Chinese medicines.

This is the tropical almond. Commonly found on tropical beaches, it is a useful emergency food.

This is the soursop: a popular tropical fruit.

The fabled durian has a repulsive smell. The pulpy flesh around the seeds is eaten. The Nuaulu erect a spear at the foot of their durian trees as a sign of ownership.

Setting Traps

The white fur and hide from the back of the ear is used to make ear ornaments worn by the men.

Smaller prey species include bats and tree frogs. Bats are the main source of food for men when they are working the sago palms in the swamp-forest of Ruatan-Nua. The bats are shot with bow and arrow. They also hunt bats in caves using makeshift ladders to reach them. In addition to the large bow and arrow made to hunt deer or pig, a smaller set is made from bamboo with very sharp bamboo arrows that is used to hunt tree frogs. A light is shone into the forest at night and the well-camouflaged frogs are shot with great accuracy.

The Nuaulu word for hunting is *yariana* and it refers only to the hunting of big game such as pythons, monitors, civets, larger bats, birds, cuscus and especially pig, deer and cassowary. The last three species are most important so they are given their own name of *peni*. Individuals and clans do have totems, but, totemic restrictions on big game are negligible. Secondary totems are respected, though not necessarily avoided as food. The division of the meat is significant. If hunting is a solitary expedition, on returning to the village the *monne*, or sacred, parts of each *peni* animal must be presented to the head of the house to which the hunter belongs. The jawbone of the pig and deer and the breastbone of the cassowary are the *monne* parts and

The bark has been partially peeled from this tree to mark the presence of a leg-hold pig snare.

Here the noose is being carefully set on a wild pig trail. It will be camouflaged so that when the unsuspecting pig treads on the trigger, the noose is tightened by a strong, springy branch.

How to make a
Spear trap

collectively are known as *penesite*, these are a gift to the house of the hunter. The meat is a gift to the person who guards the sacred house and they will then redistribute the meat to kin and others. The tail of the pig is also *monne*, but it is kept by the hunter to be made into an amulet.

When an animal is butchered, it is normally skewered on a sharpened wooden stake. After the meat has been cooked the stake is planted in the ground and the chip which was removed in making the pointed end tied to it. This is supposed to represent the re-uniting of the soul and body of the killed animal, and the whole serves as an offering to the ancestors, to return the spirit to the cosmos and thus ensure that stocks of the animals are not depleted.

Fishing
Seafish is eaten when meat is not available, but it is generally bought from the coastal people. Freshwater fish are caught by damming and poisoning part of a stream, the fish are stunned by the poison and collected. The women make intricate conical baskets from a single piece of bamboo and use them for collecting freshwater shrimp, crayfish and crabs along the banks of streams and larger rivers.

Food Preparation
Traditionally the women prepare all the food. Hot rocks are used for some occasions though most women now use metal pots.

The traditional method of cooking is in a green bamboo tube. A piece of bamboo is cut with an internode at one end. The food, which may be chopped up pieces of cuscus or shrimp are

1 This sign indicates that a spear trap has been set nearby. It is a very necessary safety feature for these traps can prove lethal to both animals and humans alike.

2 A springy arm is set between two uprights so that it can be pulled back away from the animal trail but retaining a strong springy energy back towards the trail.

3 A sharp spear is lashed to the spring arm with rattans and set on a guide to run towards the trail at the correct height.

4 A fine trigger is arranged to hold the arm back. The trigger is set off by a trip wire set across the animal trail.

Butchering and Skinning a Pig

Wild pork is a very popular food among the Nuaulu. A dangerous animal to trap, it takes two men to carry it through the forest.

placed inside and a bung of manioc leaves in put in the open end. The bamboo tube is placed over the fire and the meat left to steam in its own juices. When cooked, the bamboo is split and the food served in the two halves.

Traditional Medicine

There is a great deal of knowledge concerning traditional medicines, the women in particular. Much of this knowledge is semi-secret and magic and in their eyes it is only effective due to the correct magic. They understand that this knowledge is their property, it may belong to an individual clan and it is not necessarily shared.

Both the men and women use betel lime which is prepared from the nut of the betel palm which is crushed and mixed with lime and pepper. It has a mild narcotic effect and has great symbolic importance. It is used in ritual and medicine and if you are ill it may be mixed with ginger and spat over you.

Collecting

There are many other foods and materials that are collected from the forest and the immediate surroundings. Clam shells are

The body hair of the pig is burned off by wrapping the pig in dry palm fronds and igniting them.

collected and used as containers for instance to hold water when sharpening bushknives. They are a common sight placed outside the door of a Nuaulu house next to the whetstone. Smaller shells are used as domestic scrapers, spoons, to clean containers, scrape fireplaces and ladle sago porridge.

Larger crickets, grasshoppers and locusts are frequently roasted and eaten and they feature prominently in children's play. They are cooked and eaten in imitation of adult food preparation and provide hours of amusement; butterflies are attached by a thread to the thorax and kept as short-term pets. Grubs will be eaten dead or alive, cooked or uncooked. Some insects are important ingredients in hunting magic, and they are eaten in order to make the 'liver hot' so men are more efficient hunters.

Honeycomb is collected from at least one species of bee and the propolis is used as an adhesive for example in the making of ritual headdresses. Honey is used as a substitute for cane sugar and is collected as a cash crop when the price is high. Their technique used is to very slowly put a hand into the bee colony and drag out the honey, using little more protection than a bag over the hand and a covering over the eyes.

The Nuaulu people are a living example of how many indigenous groups living in the tropical belt would have existed alongside the rainforest. The Nuaulu have been in contact with people, traders and varying technologies for centuries and despite having moved from their original highland home, their basic living skills have altered little. They still remain experts at living with and making the most from the resources given to them by the forest, they are predominantly a forest people. They too have had to become involved with their country's politics in order to protect them and their lifestyle. The forests are being depleted, and discrimination against them and their religion are aspects that the Nuaulu recognise and need to protect themselves against.

After singeing, the pig is butchered into small pieces and stuffed into bamboo tubes for cooking.

The bamboo tubes are cooked like this on a fierce fire for over an hour. They are turned occasionally, using fire tongs made by heating and bending part of a palm frond.

Conclusion

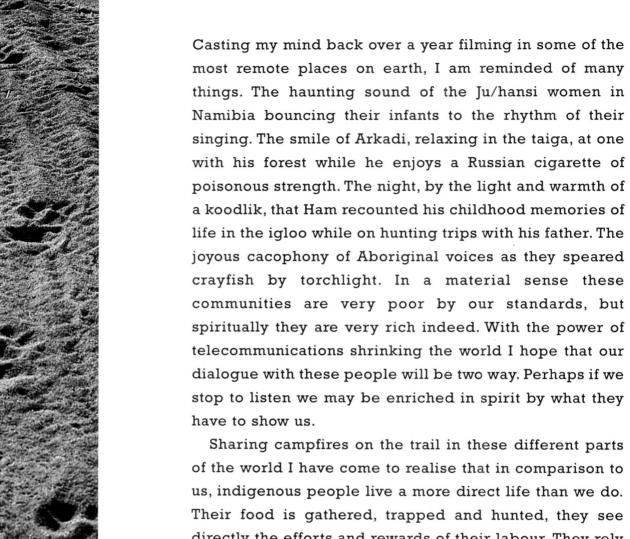

Casting my mind back over a year filming in some of the most remote places on earth, I am reminded of many things. The haunting sound of the Ju/hansi women in Namibia bouncing their infants to the rhythm of their singing. The smile of Arkadi, relaxing in the taiga, at one with his forest while he enjoys a Russian cigarette of poisonous strength. The night, by the light and warmth of a koodlik, that Ham recounted his childhood memories of life in the igloo while on hunting trips with his father. The joyous cacophony of Aboriginal voices as they speared crayfish by torchlight. In a material sense these communities are very poor by our standards, but spiritually they are very rich indeed. With the power of telecommunications shrinking the world I hope that our dialogue with these people will be two way. Perhaps if we stop to listen we may be enriched in spirit by what they have to show us.

Sharing campfires on the trail in these different parts of the world I have come to realise that in comparison to us, indigenous people live a more direct life than we do. Their food is gathered, trapped and hunted, they see directly the efforts and rewards of their labour. They rely upon their senses of observation and their manual dexterity. Today as we march through seasonless supermarkets picking packages from shelves, our brothers and sisters of the remote regions are foraging for wild foods. Perhaps in Africa a track is being followed in the dust; or in Samoa a clam being lifted from 20m beneath the waves of the Pacific Ocean, while in Australia it may be the beginning of the yam season signalled by the flowering of a particular plant.

In the cold climates, protective clothing has to be made. With a tailoring skill developed over thousands of years, the Evenk and Inuit women turn the pelts provided by the hunters into beautiful clothing that still functions in extremes better than our high tech alternatives. But unlike our clothing which tries to mimic their patterns, their garments can be easily repaired. Tool kits are also kept simple. In the rainforests it is the parang; in the cold lands, the axe, knife and saw. With these tools, life can be wrested from the natural surroundings. Prowess and speed with tools is greatly valued.

The survival of these skills and the knowledge that goes hand in hand with them is far from certain. Younger generations are tempted by the apparent ease of machinery and inevitable lure of the higher wages of the metropolis. But once the ability to live with the land has been lost so too is the foundation upon which these cultures are built. In Samoa, though, the chiefly system and their isolation have encouraged the continuation of old knowledge alongside the arrival of new. If one skill symbolises the old ways it is the making of fire by traditional means. In terms of survival, this is the most important skill and the first to disappear with the arrival of matches. I asked Moelagi Jackson, a talking chief in Samoa, why the fire-plough friction fire lighting technique was so necessary. She explained that 'When you have this skill you carry your fire in your mind and in your muscles'. And that for me sums up what this series was about.

Why not enrich your own mind and muscles by joining Raymond Mears and his team of instructors around their campfire on a WOODLORE wilderness bushcraft course.

For information please write to:

Woodlore
77 Dillingburgh Road
Eastbourne
East Sussex
BN20 8LS
44 (0)1323 648517